Opportunities to Improve Marine Forecasting

Committee on Opportunities to Improve
Marine Observations and Forecasting

Marine Board
Commission on Engineering and Technical Systems
National Research Council

NATIONAL ACADEMY PRESS
Washington, D.C. 1989

NOTICE: The project that is the subject of this report was approved by the Governing Board of the National Research Council, whose members are drawn from the councils of the National Academy of Sciences, the National Academy of Engineering, and the Institute of Medicine. The members of the panel responsible for the report were chosen for their special competencies and with regard for appropriate balance.

This report has been reviewed by a group other than the authors according to procedures approved by a Report Review Committee consisting of members of the National Academy of Sciences, the National Academy of Engineering, and the Institute of Medicine.

The National Academy of Sciences is a private, nonprofit, self-perpetuating society of distinguished scholars engaged in scientific and engineering research, dedicated to the furtherance of science and technology and to their use for the general welfare. Upon the authority of the charter granted to it by the Congress in 1863, the Academy has a mandate that requires it to advise the federal government on scientific and technical matters. Dr. Frank Press is president of the National Academy of Sciences.

The National Academy of Engineering was established in 1964, under the charter of the National Academy of Sciences, as a parallel organization of outstanding engineers. It is autonomous in its administration and in the selection of its members, sharing with the National Academy of Sciences the responsibility for advising the federal government. The National Academy of Engineering also sponsors engineering programs aimed at meeting national needs, encourages education and research, and recognizes the superior achievements of engineers. Dr. Robert M. White is president of the National Academy of Engineering.

The Institute of Medicine was established in 1970 by the National Academy of Sciences to secure the services of eminent members of appropriate professions in the examination of policy matters pertaining to the health of the public. The Institute acts under the responsibility given to the National Academy of Sciences by its congressional charter to be an adviser to the federal government and, upon its own initiative, to identify issues of medical care, research, and education. Dr. Samuel O. Their is president of the Institute of Medicine.

The National Research Council was organized by the National Academy of Sciences in 1916 to associate the broad community of science and technology with the Academy's purposes of furthering knowledge and advising the federal government. Functioning in accordance with general policies determined by the Academy, the Council has become the principal operating agency of both the National Academy of Sciences and the National Academy of Engineering in providing services to the government, the public, and the scientific and engineering communities. The Council is administered jointly by both Academies and the Institute of Medicine. Dr. Frank Press and Dr. Robert M. White are chairman and vice-chairman, respectively, of the National Research Council.

The program described in this report is supported by Cooperative Agreement No. 14-12-0001-30416 between the Minerals Management Service of the U.S. Department of the Interior and the National Academy of Sciences.

Library of Congress Catalog Card No. 89-62946
International Standard Book Number 0-309-04090-6

Additional copies of this report are available from:
National Academy Press
2101 Constitution Avenue, NW
Washington, DC 20418

Printed in the United States of America

COMMITTEE ON OPPORTUNITIES TO IMPROVE MARINE OBSERVATIONS AND FORECASTING

PETER R. TATRO, Chairman, Science Applications International Corporation, McLean, Virginia
KENNETH A. BLENKARN, Amoco Production Company (retired), Tulsa, Oklahoma
ROBERT T. BUSH, Universe Tankships, Inc., New York, New York
MICHAEL H. GLANTZ, National Center for Atmospheric Research, Boulder, Colorado
WILLIAM G. GORDON, New Jersey Marine Sciences Consortium, Fort Hancock, New Jersey
ROBERT E. HARING, Exxon Production Research Company, Houston, Texas
JON F. KLEIN, Consultant, Millburn, New Jersey
ALLAN R. ROBINSON, Harvard University, Cambridge, Massachusetts
KENNETH W. RUGGLES, Systems West, Inc., Carmel, California

Government Liaisons

ROBERT H. FEDEN, Office of the Oceanographer of the Navy, Washington, D.C.
JAMES S. LYNCH, National Oceanic and Atmospheric Administration, National Ocean Service, Washington, D.C.
RICHARD WAGONER, National Oceanic and Atmospheric Administration, National Weather Service, Washington, D.C.

Staff

CHARLES A. BOOKMAN, Project Director
C. LINCOLN CRANE, Jr., Staff Officer
RONALD C. TIPPER, Consultant
GLORIA B. GREEN, Project Secretary

MARINE BOARD

SIDNEY WALLACE, *Chairman*, Hill, Betts & Nash, Washington, D.C.
BRIAN J. WATT, *Vice-Chairman*, TECHSAVANT, Inc., Kingston, Texas
ROGER D. ANDERSON, Cox's Wholesale Seafood, Inc., Tampa, Florida
ROBERT G. BEA, NAE, University of California, Berkeley
JAMES M. BROADUS III, Woods Hole Oceanographic Institution, Woods Hole, Massachusetts
F. PAT DUNN, Shell Oil Company, Houston, Texas
LARRY L. GENTRY, Lockheed Advanced Marine Systems, Sunnyvale, California
DANA R. KESTER, Graduate School of Oceanography, University of Rhode Island
JUDITH KILDOW, Department of Ocean Engineering, Massachusetts Institute of Technology, Cambridge, Massachusetts
WARREN G. LEBACK, Consultant, Princeton, New Jersey
BERNARD LE MEHAUTE, University of Miamia, Florida
WILLIAM R. MURDEN, Murden Marine, Ltd., Alexandria, Virginia
EUGENE K. PENTIMONTI, American President Lines, Ltd., Oakland, California
JOSEPH D. PORRICELLI, ECO, Inc., Annapolis, Maryland
JERRY R. SCHUBEL, State University of New York, Stony Brook
RICHARD J. SEYMOUR, Scripps Institution of Oceanography, La Jolla, California
ROBERT N. STEINER, Operations, Atlantic Container Line, South Plainfield, New Jersey
BRIAN J. WATT, TECHSAVANT, Inc., Kingwood, Texas
EDWARD WENK, JR., Seattle, Washington

Staff

CHARLES A. BOOKMAN, Director
DONALD W. PERKINS, Associate Director
C. LINCOLN CRANE, JR., Staff Officer
ALEXANDER STAVOVY, Staff Officer
SUSAN GARBINI, Staff Officer
PAUL SCHOLZ, Research Fellow
DORIS C. HOLMES, Staff Associate
DELPHINE GLAZE, Administrative Secretary
AURORE BLECK, Administrative Secretary
GLORIA B. GREEN, Project Secretary
CARLA D. MOORE, Project Secretary

Abstract

Significant opportunities to improve marine services and products have been brought about by new observational techniques and high speed computers. Exploitation of these opportunities requires better management by increased coordination among the many sponsoring government agencies.

Hurricane forecasting was identified as an example of how effectively the present system could work when given priority and cooperation. It is mandatory that current capability be maintained.

Evolutionary improvements to the existing ocean forecasting system can be achieved by making better use of the existing data sources, by presenting the resulting analyses and forecasts more frequently to the user community with better spatial and temporal resolution, and by improving the means of dissemination.

Revolutionary improvements will require the development of global, coupled ocean-atmospheric models capable of analyzing and predicting on many time and space scales. These models are now feasible because computers of sufficient size and speed are becoming available. Atmospheric modeling is a mature science by comparison with ocean modeling; considerable effort will be required to develop the ocean models.

Both the ocean and atmospheric models will critically require significantly improved input data if they are to be successful. The United States has the opportunity to establish itself as the world leader in ocean forecasting by establishing an operational capability for forecasting critical ocean properties supported by a national operational oceanographic satellite system. Significant benefits of this system would be manifested in the areas of transportation, ocean energy development, fisheries and recreation, and coastal management.

Preface

Reliable marine observation and forecasting are essential to the safe and productive use of the sea and the coastal zone. Both commerce and the general public depend increasingly on accurate forecasts of marine conditions and weather over the oceans to prevent losses and to ensure public safety. More and more, the United States is turning to the sea for resources—energy, minerals, food, and commercial and recreational space. In some areas of the continental shelf, thousands of offshore platforms compete for space with commercial fishermen, recreational boaters, and shipping traffic. Elsewhere, fisheries management activities to 200 miles offshore have added hundreds of new vessels to the offshore fishing fleets, and increased shipping traffic has led to designated shipping lanes and vessel traffic control. Onshore, low-lying coastal communities are becoming crowded places characterized by high investment. Many new, high-risk activities are venturing into the harsh offshore environment. Public awareness of coastal crises, whether storm or pollution related, are being enhanced through media coverage.

The National Research Council, recognizing the opportunity to examine these trends and to improve safety, appointed a committee under its Marine Board to undertake an interdisciplinary assessment of the needs and expected benefits to be realized by improving the nation's ocean observation and forecasting capabilities. The committee was asked to investigate opportunities to improve marine forecasting brought about by new observational techniques and high-speed computers.

The committee was charged with

- developing clear statements of user requirements for improved observations and forecasts,

- identifying key issues and supporting facts relating to the need for and provision of improved marine observations and forecasts, and
- stimulating dialogue among all who are involved with the process of developing, providing, and using marine observations and forecasts.

The committee was composed of members with representative expertise in the fields of marine meteorology, oceanography, forecasting, and forecast dissemination and from user communities such as fishing, oil and gas extraction, and vessel operation (biographies of committee members appear in Appendix A). The principle guiding the constitution of the committee and its work, consistent with the policy of the National Research Council, was not to exclude members with potential biases that might accompany expertise vital to the study, but to seek balance and fair treatment. The committee was assisted by representatives of the National Oceanic and Atmospheric Administration (NOAA) and the U.S. Navy who were designated as liaison representatives.

The committee dealt with synoptic scale phenomena in the atmosphere and with mesoscale phenomena in the oceans. It did not address long-term, seasonal variability or other long-term phenomena. Furthermore, its inquiry was limited to physical processes; it did not address the forecasting of biological phenomena.

The committee met several times during a two-year period commencing in November 1987. Presentations were solicited from government and private organizations that provide marine forecasts. A questionnaire was developed and a survey was conducted of a representative population of the community that uses or derives significant benefit from marine forecasts and observations (Appendix B). Based on the survey results, a national meeting was convened at the Arnold and Mabel Beckman Center of the National Academy of Sciences in Irvine, California, on September 27-29, 1988. It brought together members of the user and provider communities for paper presentations, open forum discussions, and the development of working group papers on specific issues. The papers appear as part of this report. The agenda and participants of the national meeting are listed in Appendixes C and D. The results of the national meeting are documented in five working group reports that are attached to this volume (Appendixes E-I):

- Working Group 1—Wind, Wave, and Swell;
- Working Group 2—Tropical and Extratropical Storms;
- Working Group 3—Currents, Ocean Processes, and Ice;
- Working Group 4—Nearshore Forecasting; and

- Working Group 5—Collection, Reporting, Dissemination, and Display.

The committee's findings and recommendations are based on presentations made to the committee, the results of the users survey, the presentations, discussions, and deliberations of the providers and users who participated in the national meeting, and the professional experience of the committee members. The entire report has been reviewed by a group other than the authors, but only Chapters 1–3 have been subjected to the report review criteria established by the National Research Council's Report Review Committee. The workshop reports have been reviewed for factual correctness.

The committee gratefully acknowledges the generous contributions of time and information provided by the liaison representatives and their agencies and the many individuals who participated in the data-gathering process inherent to the project. Special thanks are extended to all those who communicated with the project by telephone and mail, including those who responded to the questionnaire and participated in the national meeting. Glenn Flittner of the National Marine Fisheries Service (NMFS) helped the committee clarify the scope of the project in the early months. Richard Posthumus of Sea-Land Company assisted with the design of the questionnaire and in identifying potential recipients in the maritime industry. Walter Pereyra of Profish International provided valuable insight concerning the environmental forecasting needs of the Northwest fisheries. Cathy Beech of Science Applications International Corporation provided administrative assistance to the Marine Board staff. The extraordinary cooperation and interest in the committee's work of so many knowledgeable individuals were both gratifying and essential.

Contents

EXECUTIVE SUMMARY xiii
   Findings, xv
   Recommendations, xix

1 THE MARINE OBSERVING AND FORECASTING SYSTEM ... 1
   Observing Systems, 1
   Data Collection, 5
   Global Weather and Ocean Prediction, 6
   Tailored Marine Forecasting, 8
   Product Dissemination, 11
   Data Archival and Research and Development, 12

2 USERS OF MARINE FORECASTS 14
   Responses to Committee Survey, 15
   Workshop Description, 20
   Reconciliation of Questionnaire and Workshop Results, 22
   Economic Perspective, 24
   Expected Benefits of Forecasting Improvements, 26

3 FINDINGS AND RECOMMENDATIONS 29
   Finding 1: Improved Coordination is Needed, 29
   Finding 2: Hurricane Forecasting is Adequate and Sources of Data
      and Forecasting Techniques Should be Maintained, 31
   Finding 3: More Synoptic Data Are Needed, 32
   Finding 4: Improvements Are Needed in Resolution in Space and
      Time, and Forecast Horizon, 35
   Finding 5: Improved Dissemination Systems and Linkage to
      Navy Marine Facsimile Broadcast Are Needed, 37

Finding 6: The Need for New Systems for Forecasting Internal Ocean Weather Exists, 40

Finding 7: Efforts Are Needed to Understand and Operationally Forecast Episodic Waves and Explosive Cyclogenesis, 41

APPENDIX A	Biographies of Committee Members	44
APPENDIX B	Questionnaire and Responses	48
APPENDIX C	Workshop Participants	57
APPENDIX D	Workshop Agenda	60
APPENDIX E	Working Group 1: Wind, Wave, and Swell	64
APPENDIX F	Working Group 2: Tropical and Extratropical Storms	74
APPENDIX G	Working Group 3: Currents, Ocean Processes, and Ice	96
APPENDIX H	Working Group 4: Nearshore Forecasting	108
APPENDIX I	Working Group 5: Collection, Reporting, Dissemination, and Display	117

Executive Summary

During the past two decades the coastal areas of the United States have experienced increased population and increased commercial and recreational use both ashore and on the sea. Land values in most coastal regions have skyrocketed because of the greater demand for coastal properties. Tourism has grown into a multibillion dollar per year enterprise. Declaration of the Fisheries Management Zone and the Exclusive Economic Zone have extended U.S. commercial interests in minerals, mining, energy extraction, and fisheries farther out to sea. The majority of international trade in energy materials, bulk cargo, and finished goods continues to move by sea. Ships may be fewer in number, but they are larger in capacity; the total tonnage moving by sea continues to increase. All these activities, including the construction of private homes, waterfront structures, recreational small craft, fishing fleets, and cargo vessels, and environmental protection measures are capital intensive. Some offshore ventures, such as those of the oil and gas industry, are particularly high-value high-risk operations.

All activities at sea and on the coasts are sensitive to atmospheric and oceanic conditions. Hurricane winds, waves, and storm tides can inflict losses of life and property on a massive scale. Forecasting the intensity and track of hurricanes is of paramount importance. At the other end of the forecasting spectrum is providing information on conditions, such as ocean temperature, that can be used by the fishing industry and recreational fishing community. As use of the sea and coastal regions has grown, so has dependence on the forecast of meteorological and oceanic conditions to avoid loss, ensure the safety of life, and maximize the economic well-being of industry and commerce. In addition, the forecasting of marine conditions has secured for itself a vital role in protecting the environment both for the safe disposal of waste products and the prompt response to pollution incidents.

Recognizing these facts and further recognizing that marine forecasting is being altered by rapidly changing technology such as supercomputers and satellite-borne remote sensing systems, the National Research Council under its Marine Board commissioned a committee to undertake an interdisciplinary assessment of the needs and benefits to be realized by improving ocean data collection and forecasting. The committee was drawn together from a wide variety of backgrounds including meteorology, oceanography, numerical modeling, satellite remote sensing, forecasting, vessel operations, fisheries, and minerals extraction and from government and private sector individuals who provide today's operational forecasting capability. The committee's composition and procedures were in agreement with the guidelines of the National Research Council. The National Oceanic and Atmospheric Administration (NOAA) and the U.S. Navy provided liaison personnel to assist the committee.

The Committee on Opportunities to Improve Marine Observations and Forecasting was charged to

- undertake an interdisciplinary assessment of the needs and expected benefits associated with improving forecasts of ocean conditions, including weather over the oceans and associated technology;
- consider observations and prediction of ocean conditions including waves, currents, temperature, and ice; as well as the associated meteorological winds, air temperatures, precipitation, visibility, and cloud cover;
- consider effects on general vessel operations, navigation, search and rescue missions, marine ecosystem analysis, fisheries operations, port operations, coastal zone management, and other ocean management and operations, as appropriate;
- identify needs of users for improvement and the technical potential to meet improvement needs;
- identify potential or probable benefits;
- consider needed research and development; and
- establish a dialogue among interested parties, for example oceanographers, atmospheric scientists, remote sensing experts, and users of marine forecasts including those who conduct commercial ocean operations, and other applicable groups.

The committee developed information by reviewing literature, commissioning background papers by committee members and government agencies, conducting a survey of a wide range of users, and convening a national meeting with representation from both the provider and the user communities.

The survey questionnaire was a key element in sampling the opinions of the user community and structuring the national meeting. It asked a number of critical questions about the use of marine forecasts, sources of forecast information, method of receipt, reliability, and desired forecast features not now being received. Responses to the survey provided a good cross section of current operational users in the shipping, oil and gas, and fisheries and recreational industries and indicated that about 90 percent of commercial users of the ocean and coastal waters utilize marine weather forecasts.

From the major topics identified by the questionnaires, background papers, and a review of the literature, the format of the national meeting was developed. Presentations were invited from representatives of key governmental agencies having a function in marine forecasting and the collection of marine observations. This was followed by presentations from key user groups such as vessel operators, fisheries, dredging, and oil and gas recovery. Presentations were also made by representatives of the value-added (private forecast) community. General presentations were followed by interactive discussion among all participants. At the conclusion of the invited papers, the participants split into five working groups to address specific issues: wind, wave, and swell; tropical and extratropical storms; ice, currents, and ocean processes; nearshore forecasting; collection, reporting, dissemination, and display.

Each working group reported its preliminary findings in plenary session, followed by an open discussion. In this manner significant dialogue between all interested parties was stimulated and presented in a forum to provide input to the final committee's report. Following is a summary of the committee's findings and recommendations.

FINDINGS

1. **Improve management.** The committee found several important program areas where needed developments will be difficult or impossible without clarifications of policy and improvements in coordination.

- Who's in Charge?—All too frequently, the committee was unable to identify the person or agency clearly and singly responsible for operation of the observing and forecasting system and end user support.
- Implications of Classification of Environmental Data—The Navy is a major producer of environmental observations and analytical products, which form the technical basis of a wide range of civilian forecasts. Recent technical advances have raised the prospect that some environmental data fields and products may be classified by the Navy in the future. Planning and coordinating are essential to ensure that civilian needs continue to be met even as military requirements change.
- Public and Private Roles—There is a reasonable balance between government and private marine forecast sector products and activities.

2. Hurricane forecasting is adequate and should not be degraded. Forecasting by the National Weather Service, and user response to such forecasts, have been successful in minimizing loss of life and property damage due to hurricanes in the U.S. coastal regions. User groups are aware of the uncertainties of hurricane forecasting and generally accept the burden of "false alarm" evacuation. The potential for improvement notwithstanding, the present forecasting of tropical storms is considered satisfactory for the fishing and shipping fleets. The committee was extremely concerned that no measures, such as the withdrawal of hurricane reconnaissance aircraft by the Air Force, be accomplished in the name of economy. At present, and for the foreseeable future, no data are available that adequately substitute for aircraft data in the measurement of central pressures, wind speed, and the precise location of the storm center. Given the present low skill in hurricane track forecasting, and the potential for vast damage and loss of life that exists from any hurricane, there is no reasonable justification for removing any source of critical data on tropical storms.

3. More synoptic data are needed. The ocean, representing some 70 percent of the global surface, is a vastly undersampled region. This is a severe handicap when initializing numerical models that provide the guidance used to generate nowcast and forecast products. Even if every ship at sea could (and would) report observations in a timely manner, many global regions would be virtually devoid of data. Improvements will depend on: (1) using every technologically available resource to obtain synoptic observations, (2) not losing data available from high-seas operators, and (3) organizing to maximize the data collection and utilization efforts. The committee finds improvements are needed in each of the following areas:

- Operational Oceanographic Satellite—The nation now has no plans to field a suite of sensors tailored to measure, in an operational mode, the ocean variables deemed most critical to ocean forecasting. To have these fields measured simultaneously from an orbit optimized for synoptic forecasting transmitted to primary operational ocean modeling centers has the potential to revolutionize ocean forecasting. The present program of the National Aeronautics and Space Administration (NASA) is the nation's sole space oceanography effort. While commendable in execution of its chartered role to develop and demonstrate the technology for ocean measurements, it is not designed to function as an operational ocean satellite program. Such a program would have significant benefit to a number of commercial activities, the military establishment, and to the general public.
- Lost Data Opportunities—The committee was most concerned to learn that the current system for the timely collection of marine observations, especially from vessels at sea, is not fulfilling its potential by a wide

margin. It is estimated that 50 percent of all potential marine observations are not utilized in the numerical model runs that form the basis of forecast guidance. The reasons appear myriad, but the following could be targeted as problems warranting prompt attention: improving the number of ships that submit observations and the total number of observations from ships; improving the routine, timely delivery of vessel reports to the modeling center for the synoptic model run; and improving the overall management of the data collection effort. These and other reasons call for prompt action on the part of governmental agencies with responsibilities to collect, process, and use this data.

4. Improvements are needed in resolution in space and time, and forecast horizon. Users stated a strong desire to alter the resolution of marine forecasts either by reduction of the area covered by a specific forecast or by adding higher resolution information about weather events within a forecast area. Users also desired more frequent forecast updates, especially during dangerous periods such as storm conditions. A wide number of users would benefit from higher resolution forecasts, especially those using smaller vessels or those conducting high-risk operations in the nearshore region (0 to 50 miles). Users are often not well served by forecasts that cover general conditions over a broad region of the coastline when their operations are conducted, for example, within a 20-mile radius of a certain port within that coastal region. Operations with this type of impact include commercial and recreational fishing, recreational boating, ocean engineering, pollution abatement, dredging, and tourism. High-seas vessel operators are concerned with more details about atmospheric frontal systems, including their horizontal extent and speed and direction of travel. Large-vessel operators are primarily concerned with detailed conditions when making landfall or entering port.

5. Improved dissemination systems and linkage to navy marine facsimile broadcast are needed. No matter what improvements are made in the marine observation net, and the resultant improvements in nowcasting and forecasting skill, those results will have been for naught if the system of disseminating the information to the users cannot keep pace. As the federal agencies respond to changing technology and changing budgets, they must maintain close coordination with each other and with the user community to ensure no breakdown in the dissemination system.

The committee finds there is cause to examine NOAA Weather Radio, which is the primary carrier for ocean weather information to the fishing fleet and the recreational boater, in three areas: (1) extending the range of the signal to 50 miles, (2) timing the broadcast to treat specific marine areas at a designated time, and (3) changing broadcast content to include more fine-scale information on current and forecast weather.

While supporting NAVTEX, an international system of communication, the committee is concerned (1) that there be sufficient time allowed on the broadcast to permit the full weather forecast to be presented and (2) that there be a mechanism whereby warnings of ocean weather can be immediately transmitted. It is obvious that if forecasts are both more site specific and more frequent, the potential burden for the NAVTEX system will increase.

While marine facsimile and radio teletype are both current systems for communication of marine weather data, the committee is most concerned about maintaining the capability of the marine facsimile broadcast. This broadcast is the weather backbone of most seagoing vessels. If the Navy cancels their support of marine facsimile broadcasts, another federal agency must take up the role of providing marine facsimile support in compliance with U.S. obligations as a signatory to the Safety of Life at Sea (SOLAS) Convention.

6. The need for new systems for forecasting internal ocean weather exists. Nowcasting, as it applies to marine forecasting, is the concept of integrating satellite and conventional observations in the context of a numerical ocean model to produce the best possible description of existing conditions. Forecasting carries this concept forward to the prediction of conditions at some specified time in the future. The committee found that there exists a common national interest in and need for nowcasts and forecasts of oceanic velocity and thermal and related fields within the nearshore and adjacent deep ocean. Significant and sustainable benefits to a variety of commercial, military, and recreational oceanic activities are identifiable and are now feasible for the first time, based on existing ocean science and technology. Improved nowcasts and forecasts of internal ocean weather and related boundary processes are becoming practicable. The technology is feasible and recent advances in scientific understanding have made timely prediction realistic and accomplishable.

7. Better knowledge is needed of "bomb" storms and rogue waves. Two distinct areas of marine weather, the episodic wave and explosive cyclogenesis, present phenomena of great concern to the shipping community. The episodic (or rogue) wave presents the mariner with the unforeseen occurrence of one or trains of very large waves, much larger than expected for the sea conditions forecast. Damage or loss of cargo and potential injury or loss of personnel can result. The second is the surprise storm or explosive cyclogenesis. In this case, a storm center deepens much more rapidly than forecast, generating extreme wind and sea conditions totally unexpected by the mariner. The important operative in both events is unforecast. The physics of these phenomena are at present insufficiently understood to provide any degree of predictability with confidence. The committee finds that additional research on both these events is required.

RECOMMENDATIONS

RECOMMENDATION: Improve Management. Improved coordination of the national ocean forecasting program is of such critical importance that a review of policy should be undertaken by the administrator of NOAA and the oceanographer of the Navy. Among the specific issues of concern to the committee are

- designation of a national policy and a lead agency for an operational oceanographic satellite system;
- designation of a national policy and a lead agency for nowcasting and forecasting internal ocean weather;
- maintenance and improvement of the services provided to the civil sector; and
- maintenance of the free exchange of data and information.

RECOMMENDATION: Improve Data Collection. NOAA should make a strong effort to increase the efficient voluntary reporting of timely marine observations and to increase the number of vessels providing these important data. Automation of shipboard observation systems and the use of satellite communication links are vital to increasing the quantity and quality of marine data.

RECOMMENDATION: Improve Resolution. NOAA can and should increase the usefulness of its products where supported by present analyses and forecasts by increasing the resolution in space and time, extending the time horizon of forecasts, and increasing the frequency of issue. Future product improvements should emphasize increased resolution and meeting user needs.

RECOMMENDATION: Improve Forecast Dissemination. NOAA should develop a national strategy for marine forecast product dissemination to users. Specifically, it should

- define the role of NOAA weather radio for supporting the marine community and configure the system consistent with that role;
- structure the national plan for implementing NAVTEX so that it is responsive to the need for expanded marine forecasting service;
- provide for a full-period national marine facsimile service equivalent to the existing U.S. Navy service; and
- provide for such other services as necessary to support user needs.

RECOMMENDATION: Operational Oceanographic Satellite System.

A national program for an operational oceanographic satellite system should be established.

RECOMMENDATION: Advance the Capability for Forecasting Internal Ocean Weather. The nation should establish an operational capability for nowcasting and forecasting oceanic velocity, temperature, and related fields to support coastal and offshore operations and management. Development of these capabilities will require the establishment of an observational network in areas of high priority.

RECOMMENDATION: Research on "Bomb" Storms and Rogue Waves. The federal government should develop the capability to forecast both episodic waves and explosive cyclogenesis.

1
The Marine Observing and Forecasting System

The marine forecast delivered to the marine operator represents the culmination of a long process of data observation, data collection, weather prediction, tailored forecast preparation, and forecast dissemination. This process, depicted in Figure 1-1, is backed up by a continuing process of data archiving and research and development to provide a continuously improving product.

The smooth operation of this system involves an international effort of observation and data collection, a national effort involving activities by a large number of federal agencies, a private sector effort involving equipment and forecast services companies, and the user.

This chapter explains the steps involved in providing weather and oceanographic services to a ship at sea, and highlights the major issues involved in the process. The major point to be made at the outset is that forecast preparation is largely a serially dependent process. Before each step of the process can function efficiently, all prior actions must be completed successfully. In such a process the final product will be only as good as the weakest link in the process allows it to be. Before each step of the process can function efficiently, all prior actions must be completed successfully.

OBSERVING SYSTEMS

Observations of the atmosphere and oceans are the bread and butter of environmental operations. The oceans of the world represent 70 percent

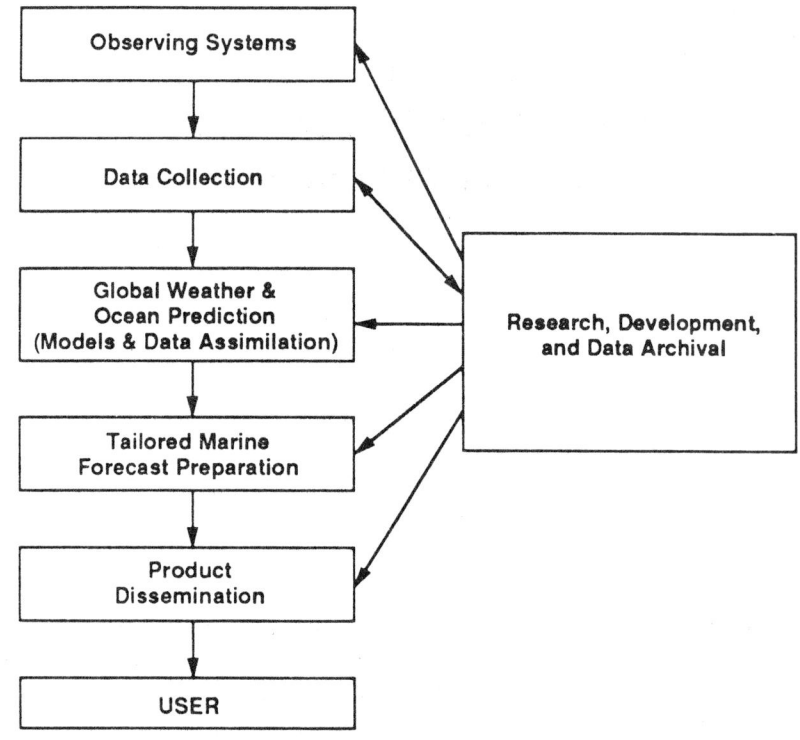

FIGURE 1-1 The marine observing and forecasting system.

of the earth's surface. In order to observe the ocean, one has to get on, over, or under the ocean with an instrument that can make a measurement. Opportunities to observe the oceans are limited to those provided by satellites and aircraft that fly over the oceans, or ships and platforms on and under the sea. One of the major challenges of improving observations at sea revolves around how ships of convenience and other user platforms can be used as sites for instruments to record and transmit vital measurements.

The National Weather Service (NWS) collects 90,000 to 95,000 worldwide marine surface weather observations monthly from cooperative weather observers aboard ships at sea. These observations include those provided by 49 countries that are recognized as contributors by the World Meteorological Organization (WMO). The U.S. Cooperative Ship Program involves 1,600 ships and is possibly the largest single national program in the world. Seventeen port meteorological officers are strategically located at NWS offices near major U.S. ports to serve as liaison to the marine community. Thirty-five operational data buoys of the National Oceanic and Atmospheric Administration (NOAA) plus 12 other prototype or special purpose buoys provide hourly meteorological and sea state observations from critical nearshore and offshore locations. In addition, 39 automated stations in the Coastal Marine Automated Network (C-MAN) report weather conditions from selected coastal sites. These marine data programs are further augmented by volunteer mariner reports (MAREPS) relayed through cooperative private coastal radio stations.

Oceanographic data include "profiles" of deep-ocean temperature and salinity that are based on measurements made by the U.S. Department of Defense (DOD) vessels, U.S. research vessels, and cooperating merchant marine and fishing vessels. Sea-surface temperatures and ocean waves are observed and reported by NOAA data buoys, U.S. Navy and domestic research vessels, and foreign commercial ships. Observations of tides, sea, and swell are also observed and reported daily. NOAA and military satellites over the ocean measure sea-surface temperature in cloud-free areas, and satellite-borne radar altimeters measure the variability of the ocean surface (ocean topography) and significant wave height.

While the numbers of observations may seem large, they are very small in terms of the size of the world's oceans. Marine weather forecasting at a resolution consistent with present computer models requires observation densities on the order of an observation for at least every 10,000 square miles of ocean area every 6 hours, augmented by higher observation densities in coastal regions where small-scale variability occurs. Within the ocean, the mesoscale variability or internal weather of the ocean occurs on space scales of tens to hundreds of kilometers and time scales of days to several weeks and requires comparable observation densities to fully define the processes. Near-surface phenomena affected by direct atmospheric forcing have more rapid variation. The present marine weather observation capability results in data densities of the order of an observation for every million square miles or less for most ocean areas, while internal ocean observations are sporadic and sparse. Moreover, adequate sampling requirements for nowcasting and forecasting of the internal structure of the ocean still need detailed determination.

Weather satellites are becoming increasingly important as the major source for surface oceanic observations. Polar-orbiting and geostationary environmental satellites can collect large volumes of weather and oceanographic data. NOAA and the DOD operate weather satellites to observe cloud cover and motion, profile vertical temperature and humidity fields in the atmosphere, measure sea-surface temperature, and portray sea and Great Lakes ice coverage.

Table 1-1 lists the present oceanographic observation satellites with

TABLE 1-1 Operational Satellites With Ocean Observing Capabilities

Satellite	Operator	Relevant Sensor
NOAA 10/11	NOAA	Clouds, surface temperature, sea ice
DMSP[a]	DOD	Clouds, ocean surface, wind speed, surface temperature, ice edge
METEOR	Soviet Union	Clouds
GOES	NOAA	Clouds, sea ice
METEOSAT	European community	Clouds
GMS	Japan	Clouds
GEOSAT	DOD	Significant wave height, sea level topography

[a] Data only available to government users.

operational capability.[1] While the United States has historically been a pioneer in satellite technology and operates many of the present weather satellites, the ocean sensing satellite systems scheduled for launch over the next decade are research missions sponsored and operated by other nations. With the right mix of instruments, the critical parameters to support ocean internal weather modeling and nowcasting could be sampled on a frequent basis. The resultant increase in observations would be at least two orders of magnitude greater than available from shipboard systems. However, the classical data handling of research missions and the management of the satellites by other nations will likely result in little impact of these satellite data on operational forecasting functions.

---

[1] Government thinking classes satellites into "research satellites" and "operational satellites." Research satellites are those satellites specifically launched to develop and test satellite technology or instrument technology. Such satellites are the sole domain of NASA. As a bureaucratic turf issue, NASA extends the view of "research satellites" to include satellites launched for the sole purpose of conducting basic research into fundamental processes. The important aspect of research satellites is that of control of the data from the satellites. Control of the data are retained wholly within the NASA family (the "principal investigators" funded by NASA) and are not publicly released except in cases of extraordinary public pressure. Furthermore, there is no commitment on the part of NASA (and often no desire) to provide data from satellites in an operationally useful time frame. By default, all other satellites are operational satellites. Operational satellites then, are those with an "operational mission" to provide public data in operationally useful time frames. DOD satellites tend to be mission-specific. They are usually launched to achieve some mission goal. The sensors and platform may be highly developmental. If the satellite mission demonstrates that it can routinely produce useful operational products, the satellite program may become operational in character.

DATA COLLECTION

For operational uses, weather data are a highly perishable commodity. Within 3 to 4 hours after observation time, processing begins to produce operational forecasts. All weather observations made over the oceans must be collected and assembled at national forecast centers within this very short time window. While observations received later than 3 to 4 hours after observation time are still useful in retrospective analysis, in forecast updates, and for use as historical record, they do not directly contribute to the accuracy of the forecast delivered to users at sea. Therefore rapid and efficient data collection and relay is an essential part of the overall system.

Ships relay their weather data through coastal stations maintained by almost all maritime countries of the world. The World Meteorological Organization (WMO) coordinates a worldwide communications system through the Global Telecommunications System (GTS) to rapidly distribute the collected observations to national and international forecast centers. The communications resources of the system, however, are maintained and operated by each country within the WMO. Therefore the efficiency of the GTS as a data collection and relay system is uneven.

The Shipboard Environmental Data Aquisition System (SEAS) program has been developed by NOAA to deliver meteorological and oceanographic data from ships operating in selected areas, accurately and quickly, to shore-based users. The SEAS equipment is portable, can be installed in a few hours, and occupies approximately 3 cubic feet of space. Using SEAS, the shipboard operator can manually or automatically enter, code, and transmit standard shipboard meteorological observations (winds, temperature, pressure, waves/swell, and ice) and oceanographic observations (subsurface temperature, salinity, and currents) via weather satellite relay. The system simplifies and streamlines the shipboard task of weather reporting and the communications relay process. The advantage to the nation is an increase in timely, accurate data from data-sparse ocean areas that will contribute to better marine forecasts to aid in safer and more economical at-sea operations.

Within the United States, the NWS, Federal Aviation Administration (FAA), and components of the DOD maintain highly efficient data collection and interchange facilities. These systems are, in general, fully responsive to the requirement for rapid data collection.

An important source for open-ocean observation data for marine nowcasting and forecasting is data from satellites. However, oceanographic satellites presently contemplated for launch are research vehicles. The operations plans for these satellites do not envision or provide for orbit selection, data sampling schemes, and the timely and expeditious processing of the satellite data in a manner useful for operational applications. Most of the data will be provided for retrospective research applications only. Therefore, while oceanographic satellites hold great promise for adequate ocean observations, their operational impact on improved forecasting will have to wait until well into the twenty-first century under present agency plans and policies.

GLOBAL WEATHER AND OCEAN PREDICTION

Global weather and ocean prediction is the process of defining the future state of the atmosphere and the oceans. The accuracy of this process depends on (1) the precision and completeness of defining the present, initial state of the domain and (2) the completeness of the simulation process (usually numerical) used to project the evolution of the domain. The completeness of the initial definition of the atmosphere and ocean area is determined by the adequacy of observations, while the completeness of the physics used to simulate the evolution of the domains depends on computer power and on understanding the physical processes in the atmosphere and in the ocean. The accuracy of the resulting product is enhanced through better observations, better knowledge about the physics of the domain, and better computers, all of which limit existing skill.

There are three major national facilities that produce global weather and internal ocean weather prediction products:

1. NOAA's National Meteorological Center (NMC) and Ocean Products Center (OPC) at Camp Springs, Maryland;

2. the U.S. Navy Fleet Numerical Oceanography Center (FNOC) at Monterey, California; and

3. the U.S. Naval Oceanographic Office at Bay St. Louis, Mississippi.

These centers operate as the main processing centers within a network of facilities involved in environmental prediction. They are equipped with extensive communication facilities and large supercomputers dedicated to operational atmospheric and ocean simulations. The NMC/OPC serves the national civil goals for weather and physical ocean prediction, while the FNOC and the Naval Oceanographic Office meet the rather specialized objectives of the U.S. Navy. However, a number of products important to marine applications are uniquely produced by the Navy centers and as such have broader value to the nation as a whole. These products are provided to NOAA for further public distribution.

Atmospheric forecasting is performed by defining the current state of the atmosphere by 3- and 6-hour pressure wind and wave analyses at the surface and 12-hour analyses at selected levels above and below the ocean surface. These products are produced by a mix of computerized numerical techniques and human operations to develop a three-dimensional picture of present and future weather and ocean conditions. Computerized forecasts are then run using simulation models to project the future state of the atmosphere.

In the ocean, a newly emerging capability within the Navy is the forecasting capability at the Operational Oceanography Center at the Naval Oceanographic Office. This center routinely produces high-resolution, local-scale analysis and forecasts of ocean mesoscale phenomena. This new capability reflects the transition of a national ocean prediction capability from research to operational application.

NMC and FNOC transmit these analyses and forecasts to field offices throughout the nation and the world and to other users, both domestic and international, for the preparation of short- and medium- range forecasts.

NOAA's Center for Ocean Analysis and Prediction

NOAA has established a new facility at Monterey, California that will collocate at FNOC. A number of related specialties and disciplines are being assembled in order to focus Navy capabilities for providing oceanic products and services for meeting the civilian objectives of the Climate and Global Change Program and for addressing coastal ocean issues. The FNOC, while producing highly specialized oceanographic products for Navy use, produces unique products for marine applications and has extensive capability to serve a broader national requirement.

The principal purpose for the center is to support NOAA line components in the performance of their mission to deliver oceanic products and services. Its particular focus will be to develop and provide products that describe and predict the variability of biological, chemical, and physical processes in the global ocean and the nation's coastal ocean. These activities link the center to NOAA's programs concerning living marine resources, habitat and coastal zone management, offshore dumping and pollution, and ocean climate processes. The center began operation in 1988. It will access the data resources available at Monterey. Types of products that NOAA intends to produce at the center during the 1990s include

- climate applications—water-level analyses/anomalies, sea ice anomalies, biological (fish count) anomalies, mass transport analyses/anomalies, global ocean flux analyses/anomalies, ocean circulation anomalies, daily global and regional MLD analyses/anomalies, and upper-ocean heat content anomalies;
- coastal environmental applications—water-level analyses/anomalies, biological analyses and assessments, chemical analyses/anomalies, ocean temperature/salinity analyses/anomalies, coastal ocean front and current analyses, mass transport analyses/anomalies, and pollution dispersion forecasts.

TAILORED MARINE FORECASTING

The process of converting a global prediction into a specific statement about future weather and ocean conditions for a particular region is called forecasting. The forecasting process involves additional computer simulations, the application of data and customer requirements to the predictions, and the use of skilled reasoning by forecasters.

There are two classes of forecasts generally issued by forecasters: (1) general public forecasts and warnings and (2) customer or requirement-specific forecasts and warnings. The marine high seas and coastal warnings and forecasts issued by NOAA are examples of a public forecast, while those issued by Navy forecasters to Navy customers or by private weather service forecasters for their customers are examples of a customer-specific or tailored forecast service.

U.S. Government Forecasts

The NWS of NOAA has the principal responsibility for the plans and operations of the nation's basic weather services and certain specific applied services. The basic mission of NWS is to help ensure the safety and welfare of the general public as it is affected by weather. In support of this mission the NWS issues warnings and forecasts of weather and ocean conditions.

NWS provides two broad types of services: (1) real-time operation-oriented services and (2) technical, advisory, and other support services.

The three principal real-time operational services are (1) the measurement and description of the meteorological and hydrological conditions that prevail; (2) the prediction of the future state of these conditions; and (3) the warnings of specific conditions that threaten life, property, and the conduct of business.

The NWS *forecasting services* involve the prediction of the future state of these same measurements for various time periods. The content of the forecasts is influenced by the interests and the requirements of the various groups of users. Forecasts are issued on a regular basis.

The *warning services* are keyed to the occurrence of specific events or conditions, such as hurricanes or tornados.

The additional *advisory and supporting services* of the NWS include assistance through the Voluntary Cooperation Program of the WMO.

Fifty-two field offices prepare and issue medium- and small-scale forecasts, weather watches, and warnings; they also acquire meteorological data. There is essentially one field office per state. Two hundred twelve local Weather Service Offices issue small-scale forecasts and weather warnings.

The National Hurricane Center in Miami, Florida, issues advisories, watches, and warnings describing the current and future location, intensity, and movement of hurricanes and other tropical storms threatening the continental United States.

To meet a significant need for an integrated analysis and applications system to support NOAA's coastal estuarine environmental, fisheries, and global change activities, several NOAA programs are being focused at developing new, and integrating existing, marine analyses and forecast products. NOAA is designing and developing the Interactive Marine Analysis and Forecast System (IMAFS) to meet specific requirements for coastal ocean programs.

IMAFS will store, process, and display conventional observations, gridded fields, digital satellite data, and climatologies; permit the overlay of multiple data and products sets; and include interactive applications capabilities. The communications capabilities of the system are being designed to provide a wide-area network, which will make accessible data and generalized large-scale products from appropriate central data bases. Ports at the IMAFS sites will permit the use of local-area networks.

Together, the storage, telecommunications, and processing and display capabilities of IMAFS will allow NOAA to apply integrated oceanic, atmospheric, and biological data sets to

- fisheries applications—pelagic and coastal fisheries management, environmental and habitat conservation, and resource assessment;
- coastal environmental applications—coastal zone and estuarine studies, pollution monitoring and control, and habitat monitoring and control;
- climate and global change applications—ENSO (El Niño, southern oscillation), global change, and atmospheric mass transport; and
- other—applied research and data quality control.

The Navy has a forecasting field office structure to serve Navy needs, several of which impact the broad national marine forecast and warning capability. Three regional Naval Oceanography Centers—the Naval Western Oceanography Center (NAVWESTOCEANCEN) at Pearl Harbor, Hawaii, the Naval Eastern Oceanography Center (NAVEASTOCEANCEN) at Norfolk, Virginia, and the Naval Polar Oceanography Center (NAVPOLAROCEANCEN) at Suitland, Maryland—are assigned broad fleet support services and related matters within their specific geographical areas of responsibility. NAVWESTOCEANCEN is responsible for the Pacific and Indian Ocean areas; NAVEASTOCEANCEN for the Atlantic and Mediterranean Sea areas; and NAVPOLAROCEANCEN prepares forecasts for the Arctic and Antarctic areas. The NAVPOLAROCEANCEN also contains the Joint Ice Center, a NOAA-Navy polar sea ice forecasting center meeting national needs. All of these centers utilize basic and applied numerical products from the FNOC. Products produced by the centers support environmental broadcasts and provide tailored support in response to specific requests from the operating forces.

Two Naval Oceanography Command Centers (NAVOCEANCOM-CENs) are located at Rota, Spain, and on the island of Guam. NAVOCEANCOMCEN Rota assists NAVWESTOCEANCEN with provision of environmental services in the western Pacific and the Indian Ocean areas. Both of these centers provide fleet environmental broadcasts and tailored support in a manner similar to the regional centers. NAVOCEANCOMCEN Guam has an additional responsibility for operation of the Joint Typhoon Warning Center (with the Air Weather Service of the U.S. Air Force), providing tropical warnings to the Air Force and issuing tropical cyclone warnings to U.S. interests in the western Pacific and Indian oceans.

Private Sector Forecasting

A major and growing sector of the national weather and ocean forecasting capability is the private weather forecasting industry. Private companies provide customized forecasts and other weather services to clients for a fee.

The employment of private sector weather forecasters in the United States is not new. However, prior to World War II, there were only a few private meteorologists. At that time most private sector meteorologists were employed by industry, primarily the airline industry. Other users were shipping companies, insurance companies, and public utilities.

Since the end of World War II, however, the private weather service industry has grown. Today, there are about 100 companies that provide weather services as a commercial product. The majority of these companies are small, with 5 to 10 employees, but some are sizable corporations with staffs of several hundred employees. The gross sales for the industry are estimated at about $150 million annually.

As the industry has grown, private weather services have begun furnishing routine forecast and weather services to the general public. This was made possible by the electronic media. In all major metropolitan areas, most of the weather forecasts distributed to the general public through local television and radio stations are prepared by private meteorologists, who tailor federally provided observations, global predictions, and warnings.

The role of the private sector weather industry is expected to increase even more rapidly during the next 10 to 15 years. There is a growing demand by the general public for improvement in both the quality and quantity of weather services. Where there has been a clearly identified demand for improved service and the ability to generate revenue for providing a service, the private sector has been highly responsive to providing effective and efficient services.

The emergence of a strong private weather forecasting industry has brought the issues of public and private roles in weather services into sharp focus. Debate continues over how the public interest at large is best served through defining the roles of NOAA's National Weather and National Ocean services, DOD's meteorological and oceanographic organizations, and the private industry service companies. The issues become particularly complex in the marine environment because of the difficulties of obtaining marine observations, disseminating forecasts to ships and other marine users, and meeting the need for private weather services to be economically viable operations.

PRODUCT DISSEMINATION

Dissemination is the process of delivering observations and forecast products to the end user, the marine operator. The process is complicated by the fact that the marine user's needs are diverse. Moreover, many of the marine users are remote from conventional shore-based communications, such as telephones and data links, and therefore depend on satellite or high-frequency radio communications. To this end, most major maritime countries maintain comprehensive marine weather broadcast capabilities to support their national maritime interests. The U.S. civil marine forecast dissemination capability ranks with those of the lesser-developed nations of the world, and continues to deteriorate on a year-by-year basis.

Large, well-financed marine operators, such as the U.S. Navy, offshore oil and mineral exploration and production operators, and major shipping lines, provide their own in-house dissemination systems to support their unique needs. It is the large number of open-ocean fishermen, tug and barge operators, pleasure boaters, and coastal operators that are poorly served by the system. In terms of raw numbers of users, this aggregation of smaller users represents the majority of the total users.

For phone-based users, NOAA operates a highly developed product dissemination system that includes

- direct radio broadcasts to the public through the very high frequency (VHF) NOAA Weather Radio system;
- facsimile broadcasts to government and nongovernment users;
- automatic telephone answering devices operated by telephone companies that directly give the public weather information furnished by NWS stations;
- direct NWS-to-the-public telephones, including automatic answering devices at NWS field offices and personalized services for public civil preparedness officials;
- cooperative "hotline" telephone answering services that provide access to the latest hurricane advisories on a fee-per-call basis;

- special interfaces to the communications systems of the agencies; for example, Federal Aviation Administration (FAA) and Coast Guard networks, civil defense systems, and systems operated by private companies; and
- a "family" of services for high-volume data users accessed in Washington, D.C., including the Public Product Service channel, Domestic Data Service, International Data Service, and Numerical Product Service.

Unfortunately, few of these services adequately serve the marine user on the seas. The direct public broadcasts over the NOAA Weather Radio support the coastal marine operator to the extent of the system's limited range and to the extent of the marine forecast time provided on the broadcast. The radio facsimile and radio teletype broadcasts are scheduled into limited time slots on Coast Guard marine frequencies, resulting in brief information transmissions that can be captured only by an alert marine operator. The landline services either directly or indirectly serve the casual pleasure boater or day sailer, but do not support the extended coastal or offshore operator. As a consequence, the U.S. open-ocean operator uses the services of other countries, if possible, and the services of the U.S. Navy full-period marine weather facsimile broadcasts issued from the regional Naval Oceanography Centers.

DATA ARCHIVAL AND RESEARCH AND DEVELOPMENT

The final arm of the provider picture is data archival and research and development. Data archival is fundamental to the support of the marine forecasting operations and its supporting research and development. All data that are received are screened for quality and retained in historical data files for future scientific and engineering applications. NOAA maintains the national atmosphere and ocean data archives in the National Oceanographic Data Center (NODC) for the oceans and the National Climatic Center (NCC) for the atmosphere. Satellite data are archived at the NCC.

Strong research and development is fundamental to improving scientific capabilities and for providing the opportunities for the next generation, whose creativity and inspired management will implement services of tomorrow. Strong university programs not only provide the improved basic understanding needed to define and predict the ocean environment more accurately; but more important, they provide the cadre of trained scientific staff needed to staff and operate the entire environmental services system.

Research and development to improve marine forecasting is a broadly based program. From a societal sense, it involves nearly every department within government, universities, and private industry. The dominant activities are those of the National Science Foundation, the U.S. Department of Defense, the U.S. Department of Energy, and the National Oceanic and Atmospheric Administration. From a technological sense, applicable research programs involve a broad spectrum of technologies, including atmospheric and oceanic physics, computers and computing methodologies, mathematical modeling, measurement and instrumentation, and many more basic studies.

There has been considerable progress in recent years in research and development. Understanding of phenomenology of the ocean has progressed to the point where the first, basic set of internal ocean forecast models can be operationally employed, allowing the transition of ocean forecasting into a viable operational capability. In meteorology and in the marine boundary layer, forecast models of the atmosphere and ocean waves have become more precise and more accurate. Technology programs have advanced in the areas of super- and micro-computers and in satellite remote sensing, creating new opportunities for advancing the operational capabilities for improved marine services.

2
Users of Marine Forecasts

The cornerstone of the operation of the committee was the detailed examination of the needs of the community using or deriving significant benefit from forecasts of the marine environment. While the providing federal agencies are called upon to give lengthy account of their programs to Congress and the Administration on a yearly or biannual basis, the user has little or no forum. The user community is broadly based and represents diverse segments of the public and private industry. As with many federal services, users of marine environmental information often take what they get with little leverage in directing or participating in any change.

The committee sought to alter that condition by actively seeking the views of the user community. Two primary mechanisms were used for this. First, a questionnaire was sent to a wide number of representatives of user communities. Second, a national workshop was conducted with invited papers and participants. The workshop was organized based on the returns of the questionnaire. Findings of the participants in the workshop were spelled out in five working group reports, found in Appendixes E–I. While it obviously was not possible to sample the views of all users of marine weather and ocean weather information, the committee believes that a broadly based representative sample has been taken and the major issues have been identified.

The early steps in the year-long process of identifying user needs were highly conditioned to well-known needs and traditional services. Later, especially at the workshop, users became aware of promising new technology applications, especially the forecasting of internal ocean weather.

TABLE 2-1 Questionnaire Response Categories

User Categories	Responses

Shipping
  tankships, bulk carriers
  cargo, containerships
  coastal tug, barges
	33

Oil and gas
  offshore oil operators,
  offshore supply, transport
  construction, drilling, support
  marine helicopters
	21

Fisheries and recreation
  commercial fishermen,
  recreational boating,
  oceanographic and fisheries research
	43

Others	4
Total	101

This growth in users' understanding of the opportunities associated with advances in marine forecasting is chronicled in this chapter.

RESPONSES TO COMMITTEE SURVEY

In March 1988, 415 questionnaires were sent to a wide variety of users of marine forecast products. The intent was to obtain as broad a distribution as possible, both geographically and functionally. Several members of the committee participated in the process of selecting individuals and groups to whom the questionnaire was addressed. The questionnaire and statistical compilations of the responses are shown in Appendix B. Approximately 100 responses were received, showing that 90 percent of commercial users of the ocean and coastal waters utilize marine weather forecasts. It is evident that virtually all the responses can be conveniently grouped for analysis purposes into three major user categories, as summarized in Table 2-1.

The oil and gas responses are heavily biased toward Gulf of Mexico operations; responses from other geographic areas are very sparse. On the other hand, fisheries and recreation responses are widely distributed geographically.

The following discussions provide a summary of the responses of these major groups by question.

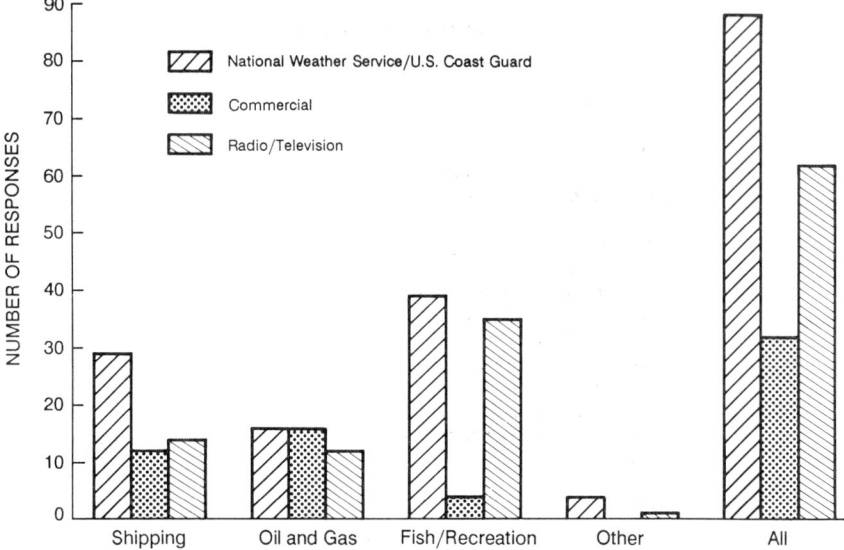

FIGURE 2-1 Sources of marine weather forecast.

Question 1: Are You Using Marine Forecasting Services?

This question was designed to establish whether or not marine forecasting services are used, and if so, what are the sources of weather and ocean conditions information, the methods of receiving the forecast, and the perceived reliability of the information. Overall utilization of marine forecasting services was high, about 90 percent of the respondents use them.

The three major sources of marine forecasts identified were U.S. Coast Guard and National Weather Service (NWS), commercial services, and radio and television. The utilization distribution for these services was not uniform among the user categories. For example, the fisheries and recreation users made the least use of commercial services and the most use of radio and television. This information is displayed graphically in Figure 2-1.

The method of receiving weather and ocean conditions forecasts was concentrated in three areas: voice, radiotelegraphy (CW, telex), and Weatherfax. The fisheries and recreation users reported virtually no utilization of radiotelegraphy in their operations. This information is displayed graphically in Figure 2-2.

The overall rating of reliability on the numerical scale of 0 to 3 was "reasonably reliable" (2.0). There was a consistent trend among all user categories to rate commercial services somewhat ahead of NWS services,

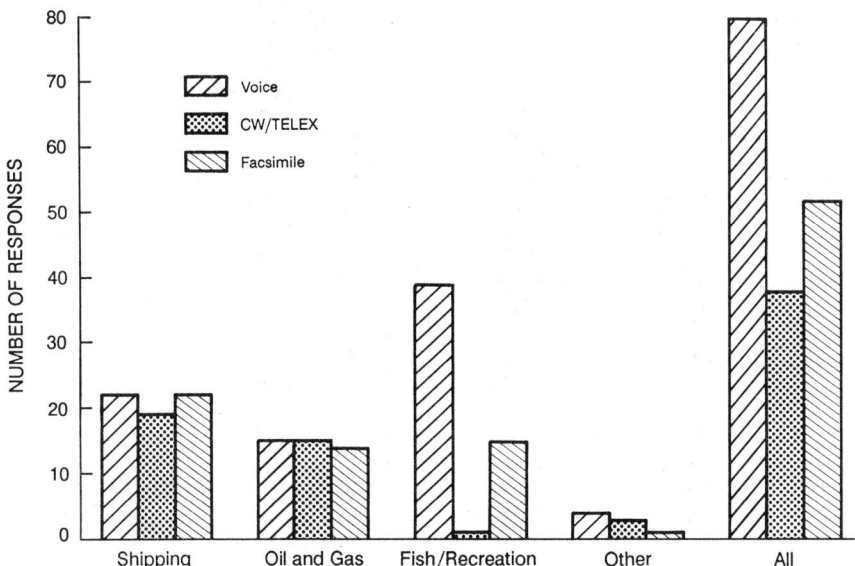

FIGURE 2-2 Method of receiving forecasts.

and both of them ahead of radio and television. This information is displayed graphically in Figure 2-3.

Question 2: What Services Beyond Those Presently Available Would You Find Useful?

This question was designed to elicit information regarding the need for services beyond those presently available. Approximately 50 percent of the questionnaires received contained such comments. These have been consolidated and are reported by major user group.

Shipping

The responses are dominated by cargo and containerships operators. The most useful services would be better access to satellite weather data through telex or high speed (1,200 Bd) modem, more frequent forecast updates, and improvements in the 12- to 18-hour forecasts. Some tankship and bulk carrier operators would find ice forecasts useful, and coastal tug and barge operators would find telephone access useful.

Oil and Gas

The responses are dominated by offshore oil operators. In the Gulf of

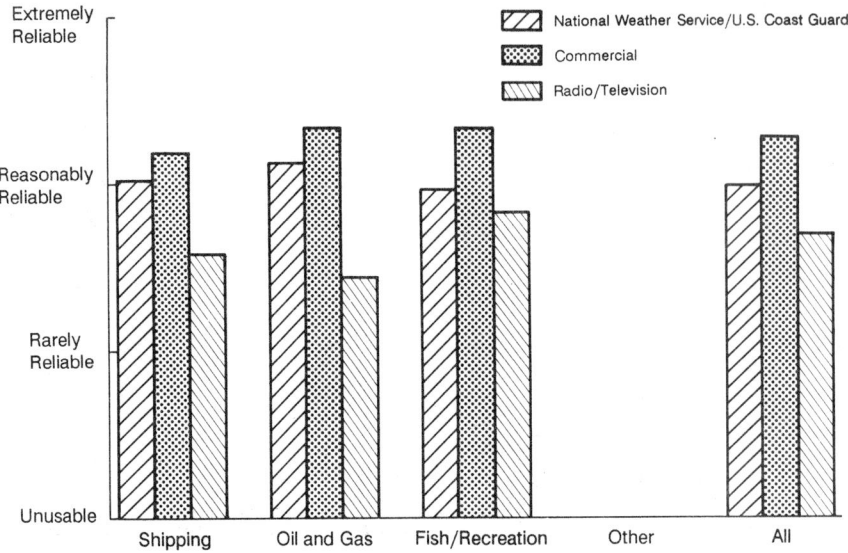

FIGURE 2-3 Reliability of marine weather forecasts.

Mexico, site-specific forecasting to at least 150 miles offshore and improved coverage of offshore VHF to 100 miles or greater would be useful. More updates during times of rapid change and the ability to track or forecast strong currents (loop or Gulf Stream) would be useful. Marine helicopter operators could use automated weather systems with real-time communication capability. On the West Coast, atmospheric stability and air pollutant concentration forecasts would be useful; offshore Alaska, ice thickness and growth forecasts and storm development offshore Siberia would be useful (when drilling operations are under way).

Fisheries and Recreation

The responses are widely spread over a variety of users. Suggestions were made for improved range of VHF coverage for "distant water" fishermen, access to offshore weather buoy information, more frequent (6-hour) updates, more satellites for weather data as well as navigation, and larger-scale area coverage. More information on ocean temperature and its variations would be useful to oceanographic and fisheries research.

Question 3: Do You Supply Observations of Marine Weather and/or Ocean Conditions to Any Organization?

This question was designed to identify how prevalent is the practice of

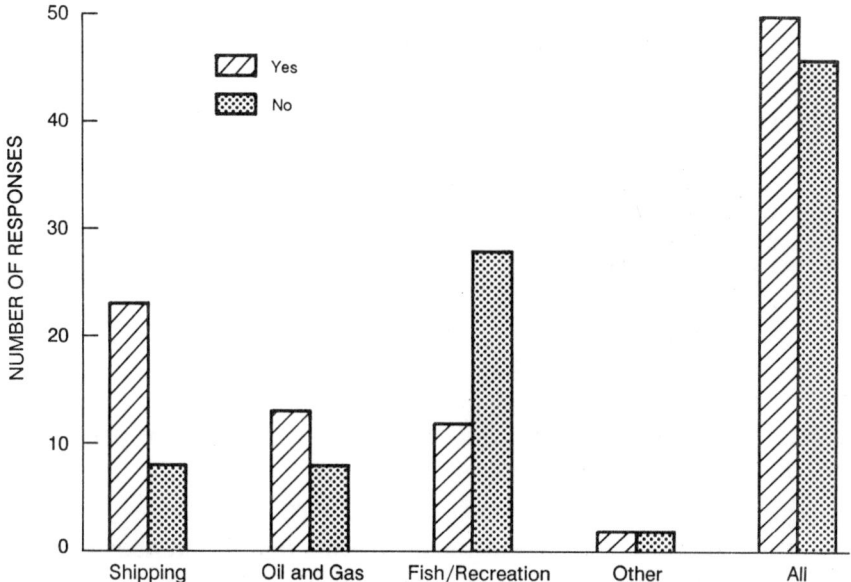

FIGURE 2-4 Supplies of marine observations.

reporting observations of marine weather conditions to other organizations. The responses were split about 50-50 in this regard. Shipping users provide the greatest proportion of marine observations and the fisheries and recreation users the least. This information is shown graphically in Figure 2-4. The recipients of these observations are quite varied and include forecast agencies as well as other operators in the immediate area.

Of those responses stating they did not supply observations of weather conditions, 50 percent said they would be willing to and 50 percent said they would not.

Question 4: Which of These Forecasted Parameters Affect Your Decisions in Marine Operations?

This question was designed to determine the priority in which forecasted parameters of marine weather and ocean conditions are important to the user's operations. A consistent pattern emerged from these responses that applies over all user groups. The summarized data focused on the five highest priority forecasted parameters for each of the several groups for which responses could be tabulated. Weighing the priorities according to the number of responses in the top five categories, the following rank order of importance to the forecast user is shown in Table 2-2. None of the user

TABLE 2-2 Importance of Forecasted Parameters to Marine Operators

Parameter	Ranking
Wind conditions	1
Tropical storm movement	2
Wave height and period	3
Swell height and period	4
Fog and visibility	5
Storm surge	6
Precipitation	7
Ice hazards	8

groups identified currents or sea temperature in the top five parameters affecting their operating decisions.

Question 5: Can You Quantify the Benefits that Would Accrue to Your Segment of the Marine Industry as a Result of Improved Marine Forecasts, and/or New Forecast Services?

Question 6: Can You Describe and Quantify Losses that Have Occurred in Your Segment of the Marine Industry as a Result of Inadequate Marine Forecasts?

These questions were designed to determine whether or not there is a basis to quantify benefits that would accrue as a result of improved marine forecasts, and to quantify losses that have occurred as a result of inadequate marine forecasts. A majority of responses indicated that benefits can be quantified (about a 3:1 margin) and that losses can be quantified (about a 2:1 margin). Descriptive examples of benefits and/or losses were provided, but no quantitative monetary value was assigned or volunteered.

WORKSHOP DESCRIPTION

On September 27–29, 1988, a national meeting was convened on "Improvements in Marine Observations and Forecasting Services: Users' Needs and Development Opportunities" at the Beckman Center of the National Academy of Sciences in Irvine, California. The national workshop was sponsored by the committee to

- develop clear statements of user requirements for improved observations and forecasts;
- identify key issues and supporting facts relating to the need for and provision of improved marine observations and forecasts; and

TABLE 2-3 Types of Participants at Committee Workshops

Category	Number of Invited Participants
Providers	
academic	3
commercial	4
government	12
military	1
Users	
coastal managers	1
fisheries and recreation	5
marine Board liaison	4
oil and gas industry	5
ports and harbors	2
shipping	7
U.S. Coast Guard	
Total	45

- stimulate dialogue among all who are involved with the process of developing, providing, and using marine observations and forecasts.

The national meeting was attended by 45 invited guests in addition to the committee members. A representative cross-section of major user groups was present, as well as a cross-section of the government, academic and private sector providers of marine forecast technology, as shown in Table 2-3. A list of participants in the national meeting is provided as Appendix C.

To stimulate discussion and establish a baseline for dialogue, the first day of the meeting was devoted to a series of papers on technical capabilities and requirements as well as the state of practice concerning observations and predictions (see Appendix D).

Open discussion and clear supported statements of fact about needs and opportunities were the hallmark of the meeting. The following issues were discussed:

- What is needed in marine observations and forecasting?
- What is available?
- If what is available is not adequate, what else is needed?
- Who should meet the needs and how?

On the second day of the meeting, all participants convened in five working groups, which met concurrently.

1. *Wind, Wave, and Swell*—Jon Klein, consultant, group leader.

2. *Tropical, Extratropical Storms*—Kenneth Blenkarn, consultant, group leader.

3. *Currents, Ocean Processes, and Ice*—Allan Robinson, Harvard University, group leader.

4. *Nearshore Forecasting*—William Gordon, New Jersey Marine Sciences Consortium, group leader.

5. *Collection, Reporting, Dissemination, and Display*—Richard Wagoner, National Weather Service, group leader.

Attendees were carefully assigned to one of the five working groups based on their interest, professional expertise, and the committee's need to maintain a balanced discussion among users and providers of marine forecasts. The goal of each working group was to extract the wisdom and perspective of users of marine forecasts on needs for improvement. Each group prepared a report that addresses the status and adequacy of marine forecasting and prediction services in the United States. Statements of fact and supporting arguments were developed in the following areas:

- specific observing and forecasting capabilities—addressed both current products and services and available data and technologies;
- needs for improvement—addressed new applications of existing technology, new technologies and data needed, and justification and costs and benefits of needed developments; and
- priorities for developments—addressed research and development, observations, operational capabilities, new technologies, and new products and services.

Membership lists for the working groups are included in Appendixes E–I. On the third day of the workshop, the leaders of each working group orally presented their results. Their reports (see Appendixes E–I) are an independent source of information for the committee and contributed materially to the development of the committee's findings.

RECONCILIATION OF QUESTIONAIRE AND WORKSHOP RESULTS

The workshop certainly achieved its objective and provided the committee with substantially more information than could be gleaned from the questionnaire responses alone. The dialogue between the providers and users of marine observations and forecasts was especially valuable and is well documented in the group reports. Some of the workshop results appear to be in conflict with or introduce topics that were not evident from the questionnaire responses. Discussion of the relationship between questionnaire responses and workshop results follows.

Question 1: Reconciliation

As expected, there was no conflict between the questionnaire and the workshop here. Participants were invited from a subset of users identified by the questionnaire. The major sources of marine forecasts identified in the questionnaire were reviewed and discussed in Working Groups 1 and 2. The method of receiving weather and ocean conditions forecasts was extensively discussed by Working Group 5, where several points were made regarding potential for improvements in the dissemination of weather products to the marine user. There was no evidence from the workshop to refute the overall assessment of "reasonably reliable" for presently available forecast services.

Question 2: Reconciliation

Very little tangible information on the need for improved services was elicited from the questionnaire. In contrast, the workshop proved to be of great benefit to the committee in this regard. For example, Working Group 1 was able to subdivide and very specifically quantify user requirements for wind, wave, and swell forecasts that exceed those presently provided. Working Group 2 took little exception to present tropical storm forecasts and warnings, but identified a significant need of the shipping industry to have better resolution of extratropical storms in the open ocean and warnings of explosive cyclogenesis. The need for warnings of episodic waves was also developed in the working groups. Working Group 3 found that there is a significant interest in and need for nowcasting and forecasting velocity, temperature, and related fields in the ocean. This information was simply not revealed by the questionnaire. The committee believes that the questionnaire responses were highly conditioned to traditionally available marine forecast products. Since ocean forecasts have been largely a military product to date, their technical feasibility, and certainly their potential availability, is unknown to the vast majority of the user community. Working Group 4 identified the special needs of the coastal or nearshore user community.

Question 3: Reconciliation

The questionnaire revealed a significant commitment to reporting marine observations by the user community. However, it also identified a substantial source of observations that are not being utilized. This theme was further emphasized in discussions of Working Group 5 concerning the lack of utilization of observations that are now provided. Working Group 2 also discussed more concentrated reporting of observations from the open ocean, particularly in the vicinity of extratropical storms.

Question 4: Reconciliation

The importance of wind, wave, and swell and tropical storm forecasts as discussed throughout the workshop certainly supports the high priority placed by the questionnaire respondents on these forecast parameters. The absence of user emphasis on currents or sea temperature in the highest priority forecasted parameters is an anomaly. It is not consistent with the findings of Working Group 3, as discussed above under question 2.

Questions 5 and 6: Reconciliation

Although a majority of questionnaire respondents claimed that benefits and losses can be quantified, the committee was unable to develop any significant documented evidence either through the workshop process or interviews with selected members of the user community. It concluded that while such quantitative benefits analyses do exist, they are rare, specific to a particular project or company, and are considered of high competitive value and hence are not generally available. If they were available, it is unlikely that they could be extrapolated to an entire segment of the user community. However, the workshop process did identify a broad range of benefits that would be achieved by improvements in the quality and method of delivery of marine weather forecasts, as well as the introduction of new forecast products and services such as those related to ocean forecasting. These benefits are summarized in the last section of this chapter, "Expected Benefits of Forecasting Improvements."

ECONOMIC PERSPECTIVE

The contribution of the ocean business sector to the U.S. economy can be developed in terms of value added. Data that are an extension of data traditionally used by the U.S. Department of Commerce have been developed by Pontecorvo and colleagues of Columbia University.[1] These data have been updated recently to establish estimated values for calendar year 1987.[2] They show that the ocean sector contributes about 2.6 percent to the total U.S. gross national product (GNP) of $4.527 trillion in 1987. Of the ocean sector as defined by Pontecorvo, the government accounts for approximately one-third; virtually all of this is attributable to the U.S. Navy.

---

[1] Pontecorvo, Giulio. 1989. Contribution of the ocean sector to the United States economy: Estimated values for 1987–A technical note. Mar. Technol. Soc. J. 23(2):7–14.

[2] Pontecorvo, G. et al. 1980. Contribution of the ocean sector to the United States economy. Science May 30, 1980.

TABLE 2-4

User Category	Contribution Total to U.S. GNP (billions)	
	1972	1987
Shipping	$ 3.7	$10.9
Oil and gas	2.3	11.4
Fisheries and recreation	12.2	48.0
Other	1.2	5.4
Total (commercial)	$19.4	$75.7

Source: Adapted from Pontecorvo et al. (1980); Pontecorvo (1989).

As noted in Chapter 1, the Navy Fleet Numerical Oceanography Center synthesizes marine observations and makes environmental predictions to satisfy the Navy's needs, and many of their products are provided for NOAA public distribution. Consequently the committee does not treat the government sector as a "user" in the context of Chapter 2.

In 1987, the commercial (nongovernment) ocean sector contributed about $76 billion, or 1.7 percent, to the total GNP. The magnitude of this contribution is in approximately the same scale as other well-recognized segments of the U.S. economy, such as all farms ($76 billion), all mining excluding offshore oil and gas ($74 billion), transportation other than water ($131 billion), and communications ($121 billion). Pontecorvo and his colleagues created an "ocean account" according to several criteria. On the supply side, they include extractive activities that involve extracting living or inanimate objects from the ocean and spatial activities where the primary activity uses the ocean water as a significant element in the production process and transportation over the water. On the demand side, they include demand attributable to the ocean and to geographic proximity to the ocean. Rearranging their data yields Table 2-4, which is approximately aligned according to the user community categories defined previously in this chapter.

Shipping includes marine transportation, marine cargo handling and related services, and ship and boat building. Oil and gas includes oil and gas extraction and heavy construction in the ocean sector. Fisheries and recreation includes commercial fishing and all of the retail trade, finance, insurance, and real estate associated with the ocean sector. The reason the committee includes the coastal zone infrastructure items with fisheries and recreation is because they are population intensive. Shipping and offshore oil and gas in themselves do not demand high population in the coastal zone. Less than 10 percent of the commercial ocean sector added value is attributable to dredging and miscellaneous services that are not easily assigned to the three major user categories.

Thus the committee concludes that of the $76 billion in 1987 of commercial added value, $48 billion or almost two-thirds (63 percent) is attributable directly to activities and residents in the coastal zone. Approximately $11 billion each are attributable to shipping and oil and gas activities for a total of an additional 30 percent. The magnitude of these numbers is impressive. They emphasize the contribution made by the ocean sector to the total U.S. GNP. The magnitude of the contribution to the U.S. economy of the coastal zone indicated in the economic data is consistent with the user community emphasis of this report. Of course, these activities depend in one way or another on dependable marine observations and forecasting of weather conditions over the ocean.

EXPECTED BENEFITS OF FORECASTING IMPROVEMENTS

Representatives of the user community participating in each working group helped identify benefits associated with improvements in the forecasting of marine conditions and delivery of the forecasts. Specific benefits were also identified by some of the presenters of topics at the workshop (Appendix D). Table 2-5 summarizes the way in which improvements in forecasted phenomena will benefit various segments of the user community. For purposes of this summary, forecast phenomena are naturally divided into two main categories: those related to atmospheric weather and those related to internal ocean weather. In addition to tabulating the nature of the benefit, the committee has grouped the benefits according to the three major user communities previously defined, that is, shipping, offshore oil and gas, and fisheries and recreation. An additional category of benefits identified is associated with coastal and Exclusive Economic Zone management. This fourth category reflects the importance that forecasting marine conditions has to residents of the coastal zone.

While somewhat subjective, the committee has identified specific phenomena that have a primary role in producing the benefit identified in Table 2-5 by 1's and those phenomena that have a secondary role in producing the benefit by 2's. Thus it is easy to see that improvements in each of the forecasted phenomena are expected to produce a benefit over a broad spectrum of user communities. It is also evident that the nature of the benefits are not solely economic. The specific benefits identified span safety at sea and environmental management, as well as efficiency improvements and economic loss avoidance. These factors, plus the established importance of the ocean sector to the total U.S. GNP, are ample justification for improvements in forecasting of marine conditions.

TABLE 2-5 Benefits Associated with Improvements in Forecasting Marine Conditions

Nature of Benefit	Forecasted Phenomena Atmospheric Weather-Related Wind, Wave, Swell	Storms	Internal Ocean Weather-Related Surface Currents, Ice	Internal Currents and Temperature
Shipping				
reduced passage time/fuel consumption	1	1	1	
enhanced safety of life	1	1	2	
reduced damage to ship	2	1	2	
reduced damage to cargo	2	1		2
reduced cargo discharge time (open berths)	1	1		
improved ship maintenance	2	2		
Offshore Oil and Gas				
reduced evacuation and shut-down costs		1		
improved construction efficiency	1			
improved deepwater drilling efficiency			1	1
reduced exposure to equipment damage		1		2
improved supply/personnel transport safety	2	1		

Fisheries/Recreation			
enhanced safety of life	1	1	
reduced equipment and vessel damage	1	1	
increased fisheries harvest	2	2	1
improved economic efficiency of fishing			1
increased recreational opportunities	2	2	
increased contribution to coastal economy	2	2	
Coastal and EEZ Development and Management			
optimized waste disposal	2	2	
improved timing of emergency evacuation	1	1	1
reduced costs of beach nourishment	2	2	
reduced costs of nearshore construction operations	1	1	
sewage and wastewater outfalls groins-breakwaters dredging	2	2	
improved coastal land use management	2	2	
improved oil spill cleanup	1	1	
more effect in search and rescue	1	1	

1 = Primary role in producing the benefit
2 = Secondary role in producing the benefit

3
Findings and Recommendations

The seven findings discussed below were distilled from the responses to the questionnaire, workshop presentations, open discussions, and working group reports. They represent a synthesis and consensus statement by the committee of topics that need to be addressed to improve the production and dissemination of marine forecasts in the support of safety of life and property and the enhancement of our economic use of the sea. They are prioritized by the order in which they should be implemented. Although a considerable effort was made by the committee to quantify the cost, schedule, and potential benefit that could be realized by the implementation of these recommendations, these could not be estimated with sufficient accuracy for inclusion here. Many of the recommendations can be implemented incrementally, with benefit increasing with both time and cost. The first five findings (better management, hurricane forecasting, more synoptic data, better space/time resolution, and improved broadcast services) are primarily concerned with the maintenance and incremental improvement of existing products and services. The remaining three (operational oceanographic satellite, forecast internal ocean weather, and episodic waves/explosive cyclogenesis) are necessary to move into a new generation of improved products and services.

FINDING 1: IMPROVED COORDINATION IS NEEDED

All too frequently the committee was unable to identify the person or agency clearly and singly responsible for operation of the observing/forecasting system and end user support. Throughout the workshop and in subsequent discussions by the committee, the need for improved coordination among the various agencies became evident. Nine executive level departments and nearly 40 federal agencies deal with the oceans. There is, in addition, a private sector industry of approximately $100 million annually providing forecasts and tailored products to various users.

Marine Data Collection

The committee found abundant evidence of inadequate coordination in the observation and collection of marine data. One of the working groups estimated that 25 to 30 percent of the observations collected at sea are not transmitted to shore, and a significant percentage of those observations are not used in the analysis and forecast process. Many reasons for this were related to transmission and quality control and to the lack of any responsible individual or agency.

Data Assimilation and Modeling

Data are processed and numerical models are run by both the National Weather Service and National Ocean Service components of NOAA and by the Fleet Numerical Oceanography Center of the U.S. Navy. Although the forecast products of the centers are generally available to each other, the Navy cannot depend on non-Navy sources that might disappear during times of increased tension, and NOAA must develop its own products in case the Navy classifies its products. At the present time, both agencies share unclassified data, but classified data are restricted to military use.

Product Dissemination

Nowcast and forecast data are disseminated by NOAA, Navy, Coast Guard, news media, and private services. Forecasts for the same place from two or more of these sources may not agree because they are based on different analyses and prepared by different experts. There are many offshore regions that are now covered only by Navy facsimile broadcasts. When the Navy shifts to encrypted digital transmission of its environmental information, most ships in these regions will be left with no source of information.

Private Sector Forecasts

There has been considerable debate during the past 5 years over "privatization" of forecast services. Private forecasters are concerned that the government issues, at no cost, products that they could sell to individual customers. The committee believes that the growth of the "value-added" private sector by a factor of 10 over the past decade demonstrates that there is a reasonable balance between government and private sector products.

RECOMMENDATION: Improve Management. Improved coordination of the national ocean forecasting program is of such critical importance that a review of policy should be undertaken by the administrator of NOAA and the oceanographer of the Navy. Among the specific issues of concern to the committee are

- designation of a national policy and a lead agency for an operational oceanographic satellite system;
- designation of a national policy and a lead agency for nowcasting and forecasting internal ocean weather;
- maintenance and improvement of the services provided to the civil sector; and
- maintenance of the free exchange of data and information.

These issues are discussed in detail in the following findings and recommendations. This recommendation has the highest priority because it is necessary for the successful implementation of those that follow.

FINDING 2: HURRICANE FORECASTING IS ADEQUATE AND SOURCES OF DATA AND FORECASTING TECHNIQUES SHOULD BE MAINTAINED

The hurricane (known as a typhoon in the Western Pacific Ocean and a cyclone in the Indian Ocean) is the single most feared and potentially destructive weather event at sea. These intense storms, with wind speeds that can reach well above 100 miles per hour, are well-known hazards to all forms of marine and coastal commerce. There is ample record of the loss of property and life caused by the hurricane and its winds, waves, and resultant coastal flooding. With modern satellite systems, numerical weather models, dedicated hurricane reconnaissance aircraft, and modern communications, the threat of the unannounced onslaught of these storms has been greatly reduced, especially in and around the continental United States. Accordingly, the National Weather Service has dedicated a priority effort to tracking and forecasting hurricanes from the National Hurricane Center (NHC) located in Miami, Florida. This is matched in the Pacific by a combined effort of the Navy and Air Force to man the Joint Typhoon Warning Center located in Guam and the Central Pacific Hurricane Warning Center (CPHC) in Honolulu. The NHC and the CPHC rely on conventional meteorological reports from shipping and weather centers around the globe, dedicated aircraft flights into the centers of known storms that threaten the United States, and, for detection and tracking, available satellite and radar data. This focus has been highly effective in providing nowcasts and forecasts of hurricane formation and behavior. It was the general consensus of those providing user inputs to the committee, and of the committee itself, that present efforts are adequate to the user need.[1] All parties emphasized, however, that present capabilities should not be degraded in any manner. There was general consensus that while current satellite observations are essential, the present state of sensor development and the lack of assured redundancy on orbit continue to make dedicated aircraft reconnaissance of active storm centers the only way to define central pressures, to ascertain accurate wind velocities, and often to localize the storm center when a well-defined eye is not observable from space due to clouds or darkness.

The following comments extracted from the reports of the working groups shed additional light on this situation.

> Forecasting by the National Weather Service and user response to such forecasts have been successful in minimizing loss of life due to hurricane occurrence in U.S. coastal regions. User groups are aware of the uncertainties of hurricane forecasting and generally accept the burdens of false alarm evacuation. The potential for improvements notwithstanding, the present forecasting of tropical storms by NWS is considered satisfactory by the fishing and shipping fleets. There seems to be low prospect for technical improvement in forecasting hurricane tracks until all-weather satellite remote sensing is developed and made operational.

> Federal agencies and local authorities or users should place emphasis on improving evacuation and decision making. NWS needs to assure that real-time or near-real-time, hurricane weather data gathering and distribution are maintained or even improved, such as by the installation of a coastal radar [referring to a Doppler radar or NEXRAD in the Gulf Coast region, offshore].

> There is general satisfaction with the NOAA [hurricane] products in terms of how they are handled and how people respond. There is a feeling the NWS in their hurricane watch and warning business provides what is needed. There is great concern that there be no degradation in the quality and nature of what is produced. The real-time data from aircraft flights, satellites, and buoys are being used, and there is an urgent recommendation not to cut any of these inputs.

> The NWS forecasting of hurricanes and user response has been judged a success in guarding public safety. It is important to maintain or even increase public confidence in hurricane evacuation management.

FINDING 3: MORE SYNOPTIC DATA ARE NEEDED

Typically in ocean nowcasting or forecasting, the field to be forecast

---

[1] It is important to note that the committee did not have the benefit of the views of local disaster preparedness agencies.

is underdefined. Increasingly, the amount of accurate and timely data to initialize the analysis model significantly increases the accuracy of the resulting forecast. Obtaining high-quality data on internal ocean fields and the atmosphere over the oceans and repeating this on a regular basis is the starting point for all marine forecasting. The oceans, which comprise some 70 percent of the earth's surface, are vast and remote. There are few reporting points except for islands and vessels at sea. The NWS and the Navy have extensive programs to get data reported from ships at sea. To be useful in the model runs that are the basis of forecast guidance, the observations must reach the modeling center in a timely manner. They must arrive, sometimes from far-flung locations, in time to be quality checked for errors and entered into the model. The value of satellites that can sample the entire earth's surface one or more times per day is immediately obvious. Satellites, even those with sensors greatly limited by cloud cover, can provide more data about the ocean's surface and a better areal coverage than that available from vessels steaming a limited number of great circle routes between major ports of call.

Operational Oceanographic Satellite

The nation now has no plans to field a suite of sensors tailored to measure, in an operational mode, the ocean variables deemed most critical to ocean forecasting. These sensors would include the altimeter for measuring ocean topography from which currents can be determined, the scatterometer with a primary role of measuring the wind speed and direction over the ocean and thus allowing for better estimation of wave parameters, the scanning microwave sensor to measure sea surface temperature and the presence or absence of ice and for providing another estimate of wind and waves, and the low-frequency microwave radiometer to provide a cloud-independent look at sea-surface temperature. The potential to revolutionize ocean forecasting may be realized if these fields are measured simultaneously from an orbit optimized for synoptic forecasting, and the data are transmitted to primary operational ocean modeling centers.

NASA is continuing a program to demonstrate the utility of several of these sensors. These include a high-quality altimeter with precision orbital tracking, a scatterometer, and an ocean color instrument to be flown in cooperation with industry. Flights of these instruments will be in partnership with programs of other countries to ensure a launch vehicle. NASA is also entering into a cooperative venture to obtain synthetic aperture radar data for the study of ice in polar regions. It was made clear to the committee that the NASA effort was not a program designed for provision of a near-real-time data stream to operational analysis and forecast centers. At best, the operational agencies with need for ocean remotely sensed data will continue to rely on the NOAA weather satellites, the DOD weather satellite, and whatever quasi-operational data can be gleaned from other programs, such as the extended oceanographic mission of GEOSAT and the SEAWIFS program proposed by EOSAT. While certainly beneficial and endorsed by the committee, these measures do not have the overall impact of a set of dedicated operational oceanographic sensors in orbit linked by rapid communications to the major analysis centers of NOAA and the Navy.

Lost Data Opportunities

While an operational oceanographic satellite is necessary for improved ocean forecasting, ship reports are equally necessary because only they provide data on subsurface ocean conditions. There is a surprising inefficiency in the collection of oceanic and atmospheric data from various marine platforms. The committee found that only about 50 percent of potentially available marine data reports are being operationally utilized in nowcasts and forecasts. The remaining reports are either not sent, lost in transmission, arrive too late for the model run, or contain too many errors to be useful. The provider and user representatives and the members of the committee felt that a concerted effort should be made to solve this problem. The following comments were made by the working groups.

> The present functioning of the Vessel Observation Service (VOS) is plagued by numerous problems related to quality control, timeliness of reporting, communications processing, and shipboard procedures. [For example, the committee learned that on U.S. flag ships it is common not to report observations at night because transmitting at night requires special overtime pay for radio operators.] The assessment of the disappointing effectiveness of the VOS program strongly suggests that the program suffers from a lack of nurturing.

The following deficiencies were found:

- significant loss of data within complex communication system,
- significant delay on delivery of data to forecasters,
- insufficient provision for providing unused data to forecasters,
- antiquated and slow communications, and
- insufficient use of reliable, quality controlled satellite communications.

RECOMMENDATION: Operational Oceanographic Satellite System. A national program for an operational oceanographic satellite system should be established.

RECOMMENDATION: Improve Data Collection. NOAA should make a strong effort to increase the efficient voluntary reporting of timely marine observations and to increase the number of vessels providing these important data. Automation of shipboard observation systems and the use of satellite communication links are vital to increasing the quantity and quality of marine data.

FINDING 4: IMPROVEMENTS ARE NEEDED IN RESOLUTION IN SPACE AND TIME AND FORECAST HORIZON

Many users, especially those whose use of the ocean is generally within 50 miles of shore, found that the present system of forecasting and forecast dissemination provided information that did not meet their needs. This was based on the spatial area covered by the forecast, the spatial resolution of the forecast, the time interval between forecast updates or modifications, and the forecast time horizon, that is, the future period covered by the forecast such as the 24-hour or 12-hour outlook. This situation was exacerbated for operations that were critically weather dependent. An example of this type of operation would be dredging or the operation of small pleasure craft.

Shipping

A major concern for vessel operators is the nature of extratropical storms over the high seas. Often, forecasts are for vast ocean regions, especially when large air masses dominate a region such as the Eastern North Atlantic. The vessel operators desire more specific location data on frontal systems, especially the horizontal depth of the frontal feature, the speed with which the front is progressing, and, when possible, the exact position of the center of low pressure. This information is desired on a fairly frequent interval to permit evasive action to be taken during transit.

A specific area of concern to the shipping community is the landfall region for approach to ports. Here the ship operator is interested not only in pressure systems and the associated wind and wave fields, but also visibility and, in high latitudes, ice. Topography also has an impact on wind direction and wind speed both in benign and storm situations. In general, smaller area forecasts would be beneficial to the shipping community in terms of minimizing time lost, vessel and cargo damage, and the potential loss of human life due to weather.

A review of documents recording marine losses provided to the committee by several insurance underwriters indicates that weather losses are a steady source of claims each year. While specific conclusions cannot be drawn from those documents without additional data and extensive analysis, the trend of a constant worldwide impact on shipping is clear. It should be noted that weather is a prominent factor listed in claims where total loss of the vessel is involved.

Oil and Gas Exploration and Production

Offshore oil and gas exploration and production operations often need small area and short time window forecasts. They need specific projections of sea state and wind conditions when conducting critical operations. For such highly specific operations as the evacuation by boat or helicopter of offshore oil fields that lie in the path of a severe storm, or the tow and placement of platform structures, highly site- and time-specific weather forecasts can significantly reduce the risks to both men and materials and ultimately translate into large cost benefits if done on time and without damage. A reasonably significant body of private forecasters and private forecasting service companies are key assistants to the major energy extraction companies in support of such decision making.

Fishing and Recreational Boating

This community of relatively small boat owners and operators is extremely sensitive to the local area nearshore forecast. Users commonly complain that the forecast areas are often much too large to be meaningful to an operator whose sailing radius from a port may be less than 20 miles. Forecasts that cover 100 miles or more of coastline often do not contain sufficient local detail. Local conditions may vary a great deal from the wide area forecast. General comments obtained by the committee through its survey and workshop request more detail on local wind and wave conditions, the time and speed of frontal passage, and expected conditions. Small changes in wind or wave forecast may have a real impact on this class of operator. For example, a wind speed forecast of 15 to 25 knots does not help operators who will be heading home at 15 knots, in difficulty at 20 knots, and a potential search and rescue case waiting to happen at 25 knots. These users would like additional forecasts at both ends of the time scale, more frequent forecast updates or nowcasts for local areas to support actual operations, and a better 24-hour outlook to support planning. For example, many recreational boats plan to stay in port based on a marine forecast tailored to a 100- to 200-mile coastal area. The forecast that causes numbers of recreational boaters to choose to stay in port can mean significant dollar losses to a local community, especially if a season for a particular fishery is short in duration. Overall, more frequent and more site-specific forecasts would be of extreme benefit to the fishing and recreational boating communities. The benefits would be measured in increased safety, improved utilization of resources, and a potential financial plus for the region.

Dredging and Ocean Engineering

These operations are often extremely sensitive to very local nearshore marine conditions. Dredging operations generally rely on machinery and barges that are not self-propelled and cannot avoid the onset of unexpected adverse weather. Commonly these platforms and equipment are sensitive to swell conditions in excess of 3 to 6 feet and can be carried from their moorings and set aground by strong wind and wave combinations. Knowledge of near-term weather is needed to permit adequate preparation without constant costly downtime. A typical dredging operation takes place within a 5 nautical mile radius of some given location. This highlights the site specificity of the forecasts desired. Ocean engineering activities that include such common coastal developments as construction of piers, jetties, and seawalls to laying of subsurface piping and the building of bridges all can be extremely dependent on weather for personnel safety and the prevention of equipment loss. Included under ocean engineering from this perspective is the cleanup of pollution events with the use of booms and small boats.

All of the operations discussed above could be aided by more site-and time-specific forecasts, especially in the coastal region. Additional detailed discussion may be found in the reports of Working Groups 1 and 2 (see Appendixes E and F).

RECOMMENDATION: Improve Resolution. NOAA can and should increase the usefulness of its products, where supported by present analyses and forecasts by increasing the resolution in space and time, extending the time horizon of forecasts, and increasing the frequency of issue. Future product improvements should emphasize increased resolution and meeting user needs.

FINDING 5: IMPROVED DISSEMINATION SYSTEMS AND LINKAGE TO NAVY MARINE FACSIMILE BROADCAST ARE NEEDED

The dissemination of marine weather information and the potential loss of the Navy marine facsimile broadcast was a common point of discussion by almost all vessel operators. As technology and federal budgets rapidly change, there is a strong feeling among several user communities—notably fisheries and marine transportation—that federal agencies will be setting policy and adopting new communications systems with little interaction with the users.

NOAA Weather Radio

The primary means of disseminating marine weather information to the general public, the recreational boating and fishing communities, and the commercial fishing fleet is either by commercial broadcast (radio and television) or the NOAA Weather Radio system. Many concerns about the existing NOAA radio system were driven in part by the larger number of boats using the coastal zone and the size and speed of today's boats that often can operate far offshore and remain out overnight.

The concerns can be broken into three general areas: broadcast range, broadcast timing, and broadcast content. There was nearly unanimous consensus that the range of the present system should be expanded to accommodate vessels, both commercial and private, out to a range of at least 50 miles. Much discussion arose about centering weather information for particular regions at a fixed time every hour. For example, the marine forecast for the area from river mouth x to headland y would always fall at 27 minutes after the hour. Also, significant numbers of users wanted more information on specific weather features not now included in the broadcast. For example:

- more information on fronts and frontal passage;
- elimination of divergent forecasts for the same area when there is overlap by two stations;
- more information on weather to the west, that is to say, weather coming into the forecast area; and
- more frequent updates during storms.

The NAVTEX System

The NAVTEX system is an evolving, international, direct printing information dissemination system that will be mandatory by August 1993 for cargo vessels over 300 tons and for all passenger vessels on international voyages. Its proposed range under the provisions of the Safety of Life at Sea Convention (SOLAS) is nominally out to 200 miles. NAVTEX will carry marine safety and hydrographic information. It will also provide offshore weather products for the ocean region that is nominally 60 to 200 miles offshore.[1] User community concerns about NAVTEX center on two issues:

1. Will there be sufficient time available to get out the weather forecast, especially in view of the increased demand for ocean weather and smaller area forecasts?

---

[1] Under the provisions of the SOLAS, high-seas weather information (beyond 200 miles) will be delivered over the INMARSAT system. Coastal weather information (to a nominal distance of 60 miles offshore) will be provided separately by the coastal countries.

2. Will the system be responsive enough to get out weather warnings in a near-real-time basis with no chance of a warning being omitted?

NOAA, the Coast Guard, and others involved in the evolving use of NAVTEX need to take full account of user needs and concerns in this process. The committee is concerned that NAVTEX, which will not be operational for several years in the United States, is marginally capable of handling the products available today and might well be overloaded with the addition of higher resolution products.

Marine Facsimile and Radio Teletype

The general consensus of the user community is that it is absolutely necessary to continue both of these services as essential broadcasts for marine weather and ocean weather information. The central issue in this discussion was the potential termination of some marine facsimile broadcasts by the Navy. The Navy began to use a facsimile broadcast for marine information more than 30 years ago. The broadcast was not encrypted and thus could be received by any ship at sea with the proper equipment. This broadcast became standard in the marine community and its basic scenario is copied by many other nations who have the capability and need to disseminate marine weather information.

Currently the Navy, for internal reasons, is considering terminating the facsimile broadcast. A differentiation is needed as to whether the Navy is just going to terminate broadcasting the marine facsimile information (in that case some other agency, such as NOAA, could arrange for broadcast to be accomplished if funding could be found) or if the information itself will be withdrawn from release to a civilian agency and thus to the public. The latter case would, in the view of the committee, have a serious impact on vessel safety, and should have detailed review before the marine facsimile information is withdrawn. At present there is no planned replacement for this service.

RECOMMENDATION: Improve Forecast Dissemination. NOAA should develop a national strategy for marine forecast product dissemination to users. Specifically, it should

- define the role of NOAA Weather Radio for supporting the marine community and configure the system consistent with that role;
- structure a national plan for implementing NAVTEX so that it is responsive to the need for expanded marine forecasting service;
- provide for a full-period national marine facsimile service equivalent to the existing U.S. Navy service; and

- provide for such other services as necessary to support user needs.

FINDING 6: THE NEED FOR NEW SYSTEMS FOR FORECASTING INTERNAL OCEAN WEATHER EXISTS

There exists a common national interest in, and need for, nowcasts and forecasts of oceanic velocity, thermal structure, and related fields. Significant and sustainable benefits to a variety of commercial, military, and recreational oceanic activities are identifiable and are now for the first time feasible based on existing ocean science and technology.

Nowcasting, as it applies to internal ocean weather, is a novel approach that integrates new and existing in situ and remotely sensed observations, and incorporates the data directly into realistic oceanic numerical models to define existing and future oceanic features and states. Although the need for nowcasting and forecasting of internal ocean weather was not as strongly supported by the working groups as the more obviously observable phenomena such as storms and rogue waves, it was a consensus of the committee that future improvements to ocean forecasting are critically dependent on the development of this capability. Commercial development, marine operations, and recreational use require expanded nowcasting and forecasting capability for mesoscale oceanic phenomena and related boundary processes of the U.S. coastal ocean and deep ocean.

The mesoscale phenomena (such as eddies, jets, and meanders) predominantly occur on space scales of tens to hundreds of kilometers and on time scales of days to weeks. Related boundary processes (like fronts, upwelling, advection, thermocline, and shelf-deep ocean interactions) occur on similar spatial scales, but frequently have broader temporal variability. Many of the oceanic processes that directly affect the U.S. Exclusive Economic Zone occur over and near the break of the continental shelf.

The forecast problem is of two types, involving (1) evolution via internal dynamic and (2) the response to local atmospheric forcing. Internal dynamical evolution drives the internal "weather" of the sea; the oceanic mesoscale is dynamically analogous to the atmospheric synoptic scale. Response to local atmospheric forcing occurring at and near the ocean surface (principally within the mixed layer depth) occurs at generally faster rates than that of internal dynamical evolution.

Prediction of oceanic mesoscale phenomena and related boundary processes has become feasible due to recent rapid progress in ocean science and technology. Advances in scientific knowledge of phenomena have occurred, which in turn are leading to new theories. New data are becoming increasingly available in "real-time." New and innovative platforms and instruments (land, ocean, and space based) are significantly increasing the availability of timely ocean observations.

Furthermore, predictive methodologies and techniques as well as processing capabilities are providing the essential tools to assimilate information and model the ocean. Significant advances during the past 10 to 20 years have made available supercomputers, data management and communications systems, new numerical models with realistic dynamics and real data initializations, and four-dimensional data assimilation capabilities.

Significant economic benefits can be realized by implementing nowcasting and forecasting capabilities for oceanic fields. The fishing industry can benefit substantially from reduced search time, fuel savings, increased safety, improved resource management, and possibly even by the creation of entirely new fisheries. The shipping industry can benefit from efficient and safe use of ship time, fuel savings, and avoidance of cargo damage. The offshore oil and gas and offshore construction industries can benefit from avoidance of equipment loss, unnecessary production or construction time loss, and over-engineering. The U.S. Coast Guard can benefit from an increase in the number of lives saved and amount of property recovered, as well as efficient and economic resource allocation, fuel savings, and disaster avoidance. The U.S. Navy can benefit from more effective defense measures and efficient resource allocation and utilization. Many of the users share similar needs and can identify similar economic benefits, especially within the U.S. Exclusive Economic Zone.

Improved nowcasts and forecasts of internal ocean weather and related boundary processes are well within the national means. The technology (observation, processing, and communications systems) is feasible, and recent advances in scientific understanding (phenomenology theories and numerical models) have made timely prediction realistic and accomplishable.

RECOMMENDATION: Advance the Capability for Forecasting Internal Ocean Weather. The nation should establish an operational capability for nowcasting and forecasting oceanic velocity, temperature, and related fields to support coastal and offshore operations and management. Development of these capabilities will require the establishment of an observational network in areas of high priority.

FINDING 7: EFFORTS ARE NEEDED TO UNDERSTAND AND OPERATIONALLY FORECAST EPISODIC WAVES AND EXPLOSIVE CYCLOGENESIS

Episodic Waves

Two distinct areas of marine weather were especially troubling to a significant number of users. The first area dealt with the occurrence of high waves known as episodic waves, more commonly known as "rogue" waves. These waves occur without warning and are uncommonly large for the sea states in which they are embedded. Although a body of literature exists on this phenomena dating back to the 1960s, no single mechanism or interactive series of events has been proven to be the cause of these waves or groups of waves such that the events are predictable.

The major user impact of these episodic waves is felt by ships operating on the high seas and the open-ocean tuna fishery. In the case of these smaller vessels, it has been speculated that an encounter with episodic wave events may have been the cause of the loss of entire vessels and crews.

Even for major vessels plying the world's sea lanes, the occurrence of such episodic waves is a significant problem, as noted by the following remarks made at the committee's workshop:

> Mariners consider the occurrence of [meteorological] surprises as the governing threat to safety of transoceanic passage.

> In spite of all this [weather information] we still hear cases of severe cargo damage and loss of vessels. These are usually caused by large waves. We [the shipping industry] would like reports of certain areas designated as likely high wave problem areas and if possible a degree of probability regarding what wave heights, direction, and frequencies of such large wave patterns can be expected.

> Every so often there come a series of three waves that appear out of nowhere, you're not expecting them, you're not prepared.

The consensus of the committee was that there is a body of anecdotal evidence to suggest the occurrence of waves dramatically larger than those anticipated on the basis of prevailing sea conditions. There does not appear to be technical consensus as to whether episodic waves reflect a particular physical phenomena or are instead merely a manifestation of the statistical variability within a given sea state. The cause notwithstanding, these events are perceived as a significant problem to the general mariner in terms of vessel, cargo, and financial damage and can result in the loss of life. No federal agency presently undertakes to forecast this phenomena.

Explosive Cyclogenesis

The second area of concern is the "surprise storm" referred to by scientists as explosive cyclogenesis. Explosive cyclogenesis describes extraordinary, low-pressure systems that deepen at rates of 1 millibar per hour or faster. Such storms are not well forecast by the National Weather Service. A common event in the winter months on the Eastern Seaboard is to hear that a local storm has moved on and "has passed harmlessly out to sea." With some troubling frequency these storms can suddenly strengthen, with central pressures falling much more rapidly than had been forecast, and the mariner's "surprise storm" has been born. As with episodic waves, these events are rarely forecast and can catch the vessel operator ill prepared for heavy weather, resulting in damage to or loss of the cargo, vessel, or personnel.

While the primary damage by explosive cyclogenesis is to vessels operating on the high seas, vessels and operations taking place in more nearshore regions can suffer damage from large waves propagating outward from the area of the storm. This is also the case for inshore activities such as dredging, which is extremely weather sensitive, and other forms of ocean engineering.

Explosive cyclogenesis has been the subject of study in a series of field experiments undertaken by the Navy, NOAA, and others. These initial research efforts will eventually improve our understanding of these phenomena. Research to this end needs to continue. Additional detailed discussion on explosive cyclogenesis and episodic waves can be found in the reports of Working Groups 2 and 4 (Appendixes F and H).

The user community, primarily high-seas vessel operators, would like better forecasts of these events and especially warnings, by area, when conditions exist that favor such explosive storm formation, or when a specific storm has the potential for such explosive deepening in pressure gradient that can cause onset of high winds and increasing sea state.

RECOMMENDATION: Research on "Bomb" Storms and Rogue Waves. The federal government should develop the capability to forecast both episodic waves and explosive cyclogenesis.

Appendix A
Biographies of Committee Members

PETER R. TATRO received a B.M.E. degree from the Georgia Institute of Technology, attended the Air-Ocean Environment Curriculum of the U.S. Naval Postgraduate School, Monterey, and earned a Ph.D. degree in oceanography from the Massachusetts Institute of Technology (MIT). He is corporate vice-president at Science Applications International Corporation, where he manages 200 scientists and engineers working on a wide variety of programs. These include ocean prediction computer programs, analysis of oceanographic data and archiving, and shore-based computers providing strategic and tactical acoustic predictions for operating forces. During his 20 years of Navy service he established an in-house research group at the Naval Research Laboratory, after having pioneered in the development and dissemination of numerical ocean and tactical forecasts.

KENNETH A. BLENKARN received B.A., B.S., M.S., and Ph.D. (mechanical engineering) degrees from Rice University. He retired from the position of research director, offshore technology, Amoco Production Company. He has conducted and was responsible for research, development, and testing of programs in the areas of petroleum drilling and production, ocean environmental forces, and structural design and reliability. As a researcher with Amoco, Dr. Blenkarn became closely acquainted with the procedures and problems of offshore operators, some of them dependent on the accuracy and timeliness of observations and forecasting of meteorological and oceanographic environments. He has been active in the technical committees of the American Petroleum Institute, Det norske Veritas, and the Society of Petroleum Engineers. Dr. Blenkarn is a member of the Marine

Board and has served previously on the Loads Advisory Group of the Committee on Marine Structures and on the Marine Board's Executive Committee. He is a member of the National Research Council.

ROBERT T. BUSH was born in England and educated at Southhampton University. He is general manager of operations of Universe Tankships, Inc., an affiliate of National Bulk Carriers. Universe Tankships operates large fleets of dry bulk carriers and tankers in worldwide trades. Capt. Bush served an apprenticeship with British Petroleum Tanker Company and sailed aboard various bulk carriers, tankers, general cargo ships, cable ships, and salvage tugs. He served as master of bulkers, tankers, and tugs for British, German, and U.S. owners. As marine superintendent, ashore, for National Bulk Carriers, he was responsible for port feasibility studies, construction and operation, tug and barge operations, and the berthing of large bulk carriers in exposed locations. In the mid-1970s he was senior marine advisor for Aramco at Ras Tanura, Saudi Arabia, concerned with vessel routing and channel development, offshore supply, and the lightering of project cargoes ashore, including heavy equipment lifts. From the mid-1970s until 1986, he was operations manager for Mercantile and Marine, Inc. (Texas)—general liner and bulk ship operations—and then became senior marine adviser with Phillips Petroleum Company. Duties with Phillips included ship-to-ship transfers worldwide and offshore terminal operations including the North Sea and other exposed locations enduring extreme weather and sea condtions. His responsibility for weather routing analysis was another task bearing closely on the interest of the subject NRC committee's work. Capt. Bush also represented Phillips at the Oil Companies International Marine Forum, American Institute of Merchant Shipping (AIMS), and served other national and international industry groups.

MICHAEL H. GLANTZ received a B.S. degree (metallurgical engineering) and M.A. and Ph.D. degrees (political science) from the University of Pennsylvania. He is head of the Environmental and Societal Impacts Group of the National Center for Atmospheric Research. He is also adjunct professor, Center on Agro-Meteorology and Climatology, University of Nebraska, and adjunct associate professor, University of Colorado, Department of Philosophy. Dr. Glantz is, or has been, a chairman, member, trustee, consultant editor, or participant with about 20 advisory committees, many concerned with meteorology, climate, drought, and their impacts. He has also authored or edited numerous publications on climate, weather, droughts, food production, and resulting social impacts. Dr. Glantz has received several awards and has lectured, taught, and organized conferences in his areas of expertise including climate, food, American and international politics, organization, and social impacts of policies. He has held several professional positions beginning with metallurgy and proceeding through operations research to his present work.

WILLIAM G. GORDON received a B.S. in zoology from Mount Union College and a M.S. degree in fisheries from the University of Michigan, where he continued with postgraduate studies. He is vice-president for programs, New Jersey Marine Sciences Consortium, where he handles Sea Grant relationships with the national office and member institutions, and provides all liaison functions for fisheries with federal, state, local, and industry representatives. Mr. Gordon retired from the National Oceanic and Atmospheric Administration (NOAA) in February 1987 as special assistant to the administrator, responsible for coordinating fishery activities and coordinating with other federal agencies and foreign governments. His prior position was as NOAA's assistant administrator for fisheries, managing and implementing fisheries programs to enhance industry production and marketing, and developing positions on international fisheries issues. Prior to these positions he held a succession of posts of increasing responsibility, directed at strengthening scientific research capabilities, encouraging international programs, and directing fishery management programs.

ROBERT E. HARING received B.S. and M.S. degrees in chemical engineering from Carnegie-Mellon University and a M.S. degree in mathematics from the University of Tulsa. He supervises Exxon Production Research Company's foundations and environmental analysis research, which provides technology for Exxon's worldwide offshore facility design and operations. Mr. Haring began his career with Exxon in 1956 as a chemical engineer. Since transferring to Exxon Production Research Company in 1965, he has worked in various areas of offshore research and engineering and has gained industry-wide recognition for his work in single-point moorings operations, simulation, physical oceanography, and wave forces. From 1976 to 1979, he was project manager for Exxon's Ocean Test Structure program. Throughout this work he has gained a keen appreciation for the benefits of accurate and timely marine observations and predictions. Mr. Haring is a member of the Society of Petroleum Engineers, and the Marine Technology Society.

JON F. KLEIN graduated from the U.S. Merchant Marine Academy at Kings Point with a B.S. degree and a Coast Guard license as deck officer. He received a M.S. in education and psychology from C.W. Post College. In February 1989 he accepted the position of director of international sales for COMSAT, Inc. Prior to that he was vice president, marine operations, and was responsible for the administration, management, maintenance, and repair of the Sea-Land owned and chartered fleet. This is one of the largest, most innovative and successful merchant fleets in the United

States. Sea-Land has employed weather routing services for many years to keep critical voyage schedules, save fuel, and avoid weather damage to its high-speed ships and cargoes. Recently it trained its own ship masters to analyze weather facsimile data received on board during voyages for more timely analysis and weather route planning. Mr. Klein joined Sea-Land Service in 1968 and sailed as deck officer on several vessels in the Atlantic and Pacific trades. He returned to the Merchant Marine Academy in 1972 to serve for 3 years on the staff of the Commandant of Midshipmen. He returned to Sea-Land and advanced through a number of managerial posts, in both the United States and in Europe, where he also lectured as guest professor at the World Maritime University in Malmo, Sweden.

ALLAN R. ROBINSON received his B.A., M.A., and Ph.D. degrees in physics from Harvard University. He is Gordon McKay Professor of Geophysical Fluid Dynamics, Department of Earth and Planetary Sciences of the Division of Applied Sciences, Harvard University, where his research is in oceanography, the dynamics of oceanic motions, and geophysical fluid dynamics. Professor Robinson was National Science Foundation fellow in meteorology and oceanography at Cambridge University, and has been co-chairman of the Mid-Ocean Dynamics Experiment of the International Science Council. He returned to Cambridge University as Guggenheim Fellow from 1972 to 1973, has been co-editor-in-chief of *Dynamics of Atmospheres and Oceans* and is a fellow of the American Academy of Arts and Sciences. Dr. Robinson has numerous publications and memberships and has held many leadership positions within university, government, and NRC committees and boards concerned with oceanography, atmospheric science, space systems, and advanced scientific computing. He is widely known for his ideas in ocean forecasting and the combined use of dynamic model and composite data sets. Lately he has also become well known for his expertise in interpretation of satellite remote sensing of the oceans.

KENNETH W. RUGGLES received a B.S. degree from the U.S. Naval Academy, a M.S. degree in meteorology from the U.S. Naval Postgraduate School, and a Ph.D. degree from MIT. He is president of Systems West, Inc. Dr. Ruggles has served on various national policy advisory boards and committees. From 1978 to 1986, Dr. Ruggles was vice-president, then president of Global Weather Dynamics, Inc., where he directed implementation of innovative technologies and provided general management of private weather services and message switching communications activities. He has served as consultant in a variety of meteorological and oceanographic service areas, including to the International Civil Aviation Organization and as a member of the National Blue Ribbon Panel on Information Policy Implications of Archiving Satellite Data.

Appendix B
Questionnaire and Responses

Marine Board of National Research Council - March 1988

Please return completed Questionnaire to:

C. Lincon Crane, Jr.
National Research Council
Marine Board GF-250
2101 Constitution Avenue, NW
Washington, D.C. 20418

Responding Organization_____

Nature of your organization's business_____

_____

Contact person--name, address, phone number:_____

1. Are you using marine forecasting services?   Yes   No

    a. What are your sources of weather and ocean condition observations and forecasts . .
        ___ U.S. National Weather Service/USCG Advisories
        ___ Commercial Weather Service; specify:
        ___ Local Radio - Television
        ___ Other U.S. or Foreign Sources; specify:

    b. How do you receive weather and ocean conditions forecasts?
        ___ Voice
        ___ Radiotelegraphy (CW, Telex)
        ___ Weatherfax
        ___ NAVTEX
        ___ Realtime Satellite Imagery; which satellite___

    c. From a reliability standpoint how do you rate observations and forecasts provided by . . .

	Unusable	Reliable	Reliable	Reliable
NWS/USCG	_____	_____	_____	_____
Commercial	_____	_____	_____	_____
Radio/TV	_____	_____	_____	_____
Other	_____	_____	_____	_____

    d. Additional comments concerning forecasting services (i.e., timeliness, accuracy, coverage area):

2. What services beyond those presently available would you find useful?

3. Do you supply observations of marine weather and/or ocean conditions to any organizations? ___ Yes ___ No; If yes, please specify method and to whom transmitted

   If not, would you be willing to supply these types of observations? ___ Yes ___ No.

4. Which of these forecasted parameters affect your decisions in marine operations? Please indicate priority (highest = 1; if not used, leave blank).

   a. Tropical Storm Movement ___
   b. Ice Hazards ___
   c. Surface Currents ___
   d. Sub-surface Currents ___
   e. Surface Sea Temperatures ___
   f. Sub-surface Sea Temperatures ___
   g. Fog ___
   h. Precipitation ___
   i Wave Height/Period ___
   j. Swell Height/Period ___
   k. Wind conditions ___
   l. Storm Surge ___
   m. Others; Please specify _____

5. Can you quantify the benefits that would accrue to your segment of the marine industry as a result of improved marine forecasts, and/or new forecast services?

6. Can you describe and quantify losses that have occurred in your segment of the marine industry as a result of inadequate marine forecasts?

7. Would you be willing to participate in the Workshop mentioned in the covering letter? ___ Yes ___ No

Please add any additional comments you may have:

Responses to Questionnaire #1

16-Aug-88

	Invited Users	Using Y	Using N	NWS/USCS	Commercial	Radio/TV	Other
1	U.S. Navy						
2	Military Sealift						1
3	U.S. Coast Guard	3		3		1	
4	Corps of Engineers						
5	NOAA						
6	Tankship/Bulk carriers	7	2	6	5	1	5
7	Cargo/containerships	12	4	16	5	7	1
8	Great Lakes carriers						
9	Coastal tug/barges	7	1	7	2	6	1
10	Major river tug/barges						
11	Commercial fishermen (Alaska)	5		5		4	
12	Commercial fishermen (Gulf/East coast)	9		9	1	7	
13	Commercial fishermen (ocean going)	1		1			
14	Offshore oil operators	11	2	10	9	9	1
15	Offshore supply/transport	1		1	1	1	1
16	Offshore construction/drilling/support	5	1	4	5	1	
17	Recreational boating	15	1	13	3	14	
18	Oceanographic & fisheries research	12		11		10	1
19	Marine helicopters	1		1	1	1	
20	Enforcement/environmental/oil spill	1		1			
21	Harbor/coastal pilots						
22	Port authorities						
23	O.C.I.M.F.						
		90	11	88	32	62	11

		Method of Receiving Forecast?				Reliability Scale*			
		Voice	CW/Telex	Facsimile	NAVTEX	Satellite	NWS/USCG	Commercial	Radio/TV	Other
1	U.S. Navy									
2	Military Sealift									
3	U.S. Coast Guard	3	3	1		1	2.00		2.00	
4	Corps of Engineers									
5	NOAA									
6	Tankship/Bulk carriers	5	5	9			2.30	2.30	1.30	
7	Cargo/containerships	11	10	10	2		1.93	2.16	1.60	
8	Great Lakes carriers									
9	Coastal tug/barges	6	4	3		1	2.00	2.00	1.60	
10	Major river tug/barges									
11	Commercial fishermen (Alaska)	5	1	4			1.80	1.90	1.50	
12	Commercial fishermen (Gulf/East coast)	9		5			2.00	3.00	2.00	
13	Commercial fishermen (ocean going)	1			1		2.00			
14	Offshore oil operators	11	10	10		2	2.40	2.45	1.44	
15	Offshore supply/transport	1	1	1	1	1	2.00	2.00	2.00	
16	Offshore construction/drilling/support	2	3	2		1	1.75	2.25	1.33	
17	Recreational boating	13		4		1	1.90	2.10	1.80	
18	Oceanographic & fisheries research	11	1	2		1	2.10		1.90	
19	Marine helicopters	1	1	1			1.00	2.00	1.00	
20	Enforcement/environmental/oil spill	1								
21	Harbor/coastal pilots									
22	Port authorities									
23	O.C.I.M.F.									
		80	38	52	4	8	1.99	2.28	1.70	0.00

Weight average over number of responses

0 = Unusable
1 = Barely reliable
2 = Reasonably Reliable
3 = Extremely Reliable

Responses to Questionnaire #2

12-Aug-88

	Invited Users	Number of Responses	Number of Comments/Suggestions
1	U.S. Navy		
2	Military Sealift		
3	U.S. Coast Guard	3	1
4	Corps of Engineers		
5	NOAA		
6	Tankship/Bulk carriers	9	3
7	Cargo/containerships	16	15
8	Great Lakes carriers		
9	Coastal tug/barges	8	2
10	Major river tug/barges		
11	Commercial fishermen (Alaska)	5	1
12	Commercial fishermen (Gulf/East coast)	9	4
13	Commercial fishermen (ocean going)	1	1
14	Offshore oil operators	13	11
15	Offshore supply/transport	1	1
16	Offshore construction/drilling/support	6	1
17	Recreational boating	16	3
18	Oceanographic & fisheries research	12	5
19	Marine helicopters	1	1
20	Enforcement/environmental/oil spill	1	
21	Harbor/coastal pilots		
22	Port authorities		
23	O.C.I.M.F.		
		101	49

Responses to Questionnaire #3

10-Aug-88

Invited Users	Number of Responses	Do You Supply Observations Yes	Do You Supply Observations No	If Yes, To Whom Transmitted	If No, Would You Be Willing to? Yes	If No, Would You Be Willing to? No
1 U.S. Navy						
2 Military Sealift						
3 U.S. Coast Guard	3	2	1	NWS, USN, USCG, Other vessels		1
4 Corps of Engineers						
5 NOAA						
6 Tankship/Bulk carriers	9	7	2	NOAA, NWS		2
7 Cargo/containerships	15	13	2	NWS, NAVITECH, USCG	2	
8 Great Lakes carriers						
9 Coastal tug/barges	7	3	4	NWS		4
10 Major river tug/barges						
11 Commercial fishermen (Alaska)	5	3	2	NWS	2	
12 Commercial fishermen (Gulf/East coast)	9	2	7	Other fishermen	4	2
13 Commercial fishermen (ocean going)	1		1		1	
14 Offshore oil operators	13	10	3	NWS(C-MAN), Conn svc, Net office	2	1
15 Offshore supply/transport	1	1		NOAA/NWS		
16 Offshore construction/drilling/support	6	1	5	Private forecaster	3	2
17 Recreational boating	14	4	10	MAREP, NOAA	3	7
18 Oceanographic & fisheries research	11	3	8	NOAA, other vessels	3	1
19 Marine helicopters	1	1		NWS	1	
20 Enforcement/environmental/oil spill	1		1			1
21 Harbor/coastal pilots						
22 Port authorities						
23 O.C.I.M.F.						
	96	50	46		21	21

Responses to Questionnaire #4

12-Aug-88

	Invited Users	Number of Responses	Tropical Storm	Ice Hazards	Surface Currents	PRIORITY* (HIGHEST = 1) Subsurface Currents	Surface Sea Temp	Subsurface Sea Temp
1	U.S. Navy							
2	Military Sealift							
3	U.S. Coast Guard	3	3					
4	Corps of Engineers							
5	NOAA							
6	Tankship/Bulk carriers	9	1	2				
7	Cargo/containerships	15	1					
8	Great Lakes carriers							
9	Coastal tug/barges	8	1					
10	Major river tug/barges							
11								
12	All fishing/recreation	42						
13								
14	Offshore oil operators	11	1					
15	Offshore supply/transport	1	1					
16	Offshore construction/ drilling/support	5	1					
17								
18								
19	Marine helicopters	1	1					
20								
21	Harbor/coastal pilots							
22	Port authorities							
23	O.C.I.M.F.							
	Number of top five responses		53	9	0	0	0	0
	Average priority		1.3	2.0				
	Weighted priority		2	8				
	Total responses	95						

*Top five in each user category ... composite of all responses

#4

Invited Users	Fog/Visibility	Precip	Wave NT/Period	Swell NT/Period	Wind Conditions	Storm Surge
1 U.S. Navy						
2 Military Sealift						
3 U.S. Coast Guard	1	5	3		1	
4 Corps of Engineers						
5 NOAA						
6 Tankship/Bulk carriers			4	5	2	
7 Cargo/containerships	3		2	3	3	
8 Great Lakes carriers						
9 Coastal tug/barges	3		5		2	3
10 Major river tug/barges						
11						
12 All fishing/recreation	4	5	2	3	1	5
13						
14 Offshore oil operators			2	4	2	5
15 Offshore supply/transport			1	1	1	5
16 Offshore construction/drilling/support			2	3	5	4
17						
18						
19 Marine helicopters	2				3	
20						
21 Harbor/coastal pilots						
22 Port authorities						
23 O.C.I.M.F.						
Number of top five responses	67	45	94	83	95	67
Average priority	2.6	5.0	2.6	3.2	2.2	4.4
Weighted priority	5	7	03	4	1	6
Total responses	95					

*Top five in each user category . . . composite of all responses

Responses to Questionnaire #5...#6

10-Aug-88

	Invited Users	Number of Responses	Can You Quantify Benefits? Yes	Can You Quantify Benefits? No	Can You Quantify Losses? Yes	Can You Quantify Losses? No
1	U.S. Navy					
2	Military Sealift					
3	U.S. Coast Guard	3	1	1	1	1
4	Corps of Engineers					
5	NOAA					
6	Tankship/Bulk carriers	9	8	1	4	5
7	Cargo/containerships	13	12	1	10	3
8	Great Lakes carriers					
9	Coastal tug/barges	8	3	5	2	6
10	Major river tug/barges					
11	Commercial fishermen (Alaska)	5	4	1	4	1
12	Commercial fishermen (Gulf/East coast)	9	5	3	6	2
13	Commercial fishermen (ocean going)	1	1		1	
14	Offshore oil operators	13	10	1	10	1
15	Offshore supply/transport	1	1		1	
16	Offshore construction/drilling/support	5	4		3	1
17	Recreational boating	14	9	4	9	5
18	Oceanographic & fisheries research	11	7	4	5	5
19	Marine helicopters	1	1			
20	Enforcement/environmental/oil spill	1	1			
21	Harbor/coastal pilots					
22	Port authorities					
23	O.C.I.M.F.					
		94	66	21	56	30

APPENDIX C
Workshop Participants

National Meeting on
Opportunities to Improve Marine
Observations and Forecasting
Irvine, California
September 27-29, 1988

Committee Members

Peter R. Tatro, *Chairman*, Science Applications International Corporation
Kenneth A. Blenkarn, Consultant
Robert T. Bush, Universe Tankships, Inc.
William G. Gordon, New Jersey Marine Sciences Consortium
Robert E. Haring, Exxon Production Research Company
Jon F. Klein, Sea-Land Service, Inc.
Allan R. Robinson, Harvard University
Kenneth W. Ruggles, Systems West, Inc.

Liaison Representatives

Robert H. Feden, U.S. Naval Observatory
James Lynch, National Oceanic and Atmospheric Administration (NOAA)
Richard Wagoner, National Weather Service, NOAA

Invited Participants

Richard B. Allen, Atlantic Offshore Fishermen's Association
Vincent J. Cardone, Oceanweather, Inc.

Henry Chen, Ocean Systems, Inc.
Steve Cook, National Ocean Service, NOAA
George E. Duffy, Navios Ship Agencies, Inc.
Paul Friday, National Ocean Service, NOAA
Paul Glaiber, Great Lakes Dredge and Dock Company
Gary Gridley, Conoco, Inc.
Warren W. Hader, Montauk Fishermen's Association
Glenn D. Hamilton, National Data Buoy Center, NOAA
Walter E. Hanson, International Ice Patrol, U.S. Coast Guard
Charles W. Hummer, Dredging Division, U.S. Army Corps of Engineers
Paul Jacobs, Office of Meteorology, National Weather Service, NOAA
Chester Jelesnianski, National Weather Service, NOAA
Saunders A. Jones, Puerto Rico Marine Management, Inc.
Walter Kristiansen, Amoco Transportation Company
R. Michael Laurs, National Marine Fisheries Service, NOAA
Edmond Mandin, American President Lines
Ronald D. McPherson, National Meteorological Operations Division, NOAA
Paul B. Mentz, Maritime Administration
Forrest A. Miller, Inter-American Tropical Tuna Commission
Kathleen A. Miller, National Center for Atmospheric Research
Polly Mirkovich, The Texas Shrimp Association
Christopher N. K. Mooers, Institute of Naval Oceanography
Robert J. Murray, Matson Navigation Company
David Paskausky, Research and Development Center, U.S. Coast Guard
William C. Patzert, Jet Propulsion Laboratory
David J. H. Peters, Conoco, Inc.
Admiral Richard Pittenger, Oceanographer of the Navy
D. B. Rao, National Meteorological Center, NOAA
Allan M. Reece, Shell Development Company
William S. Richardson, National Ocean Service, NOAA
Stephen K. Rinard, National Weather Service Southern Region, NOAA
Frederick H. Sharrocks, Jr., Federal Emergency Management Agency
George Sparacino, Sun Transport Company
Andrew M. Sullivan, Weather Network, Inc.
H. G. P. (Bud) Thomas, Keystone Shipping Company
Paul E. Versowsky, Chevron U.S.A., Inc.
John Vermersch, Jr., Exxon Production Research Company
Charles L. Vincent, Coastal Engineering Research Center
James W. Winchester, Association of Private Weather Related Companies
Paul Wolff, Assistant Secretary of Commerce for Oceans and Atmosphere (Retired)
Vincent Zegowitz, National Weather Service, NOAA

Marine Board Staff

Charles A. Bookman, Director
C. Lincoln Crane Jr., Staff Officer
Aurore Bleck, Administrative Secretary
Gloria B. Green, Project Secretary

Appendix D
Workshop Agenda

National Meeting on
Opportunities to Improve Marine
Observations and Forecasting
Irvine, California
September 27-29, 1988

National Academy of Sciences
Arnold and Mable Beckman Center
100 Academy Drive
Irvine, CA 92715
(719) 721-2200

Tuesday, September 27, 1988

1.0 Welcome—Auditorium	Peter R. Tatro, *Chairman*, Vice President, Technology and Science Dept., Science Applications International Corporation
2.0 Providers Overview	Kenneth W. Ruggles, President Systems West, Inc.
2.1 National Weather Service Products and Services	Richard Wagoner, NOAA, National Weather Service

2.2 U.S. Navy Products   Admiral Richard F. Pittenger,
    and Services   Oceanographer of the Navy

BREAK

2.3 NASA's Ocean Program   William C. Patzert, Jet
                                      Propulsion Laboratory

2.4 Commercial Ocean Products   Vincent J. Cardone
                                      Ocean Weather Inc.

2.5 Evolving Role of Private   James W. Winchester
    Weather Services   Association for Private
                                      Weather Related Companies

2.6 NOAA's New Coastal Prediction   Paul M. Friday, NOAA
    Facility   National Ocean Service

Lunch - Buffet (Refectory)

3.0 Users of Marine Forecasts: Robert E. Haring, Section
    Results of a Survey   Manager, Exxon Production
                                      Research Company

3.1 Commercial Shipping—Needs   Robert J. Murray
    Contributions   Matson Navigation Company

3.2 Gulf of Mexico Hurricane   David J.H. Peters
    Alert Program   Conoco Inc.

3.3 Everybody Doesn't Complain   Richard B. Allen
    About the Weather   Atlantic Offshore
                                      Fisherman's Association

3.4 Weather Forecasting for the   Saunders A. Jones
    Man on the Bridge   Puerto Rico Marine Management

BREAK

3.5 Forecasting System for   Allen M. Reece
    Deepwater Drilling   Shell Development Co.

3.6	Weather Forecasting and the Dredging Industry	Paul H. Glaiber Great Lakes Dredge and Dock Company
3.7	Needs of Estuarine and Coastal Recreational Boat Operators	Warren W. Hader Montauk Fishermen's Association
3.8	U.S. Coast Guard Activities Involving Marine Observations and Forecasting	Walter E. Hanson U.S. Coast Guard

ADJOURN

Dinner: Irvine Hilton
and Towers (Imperial Room)
"Marine Forecasting:
Perspective on Future Developments"

Guest Speaker: Paul Wolff, Assistant Secretary of Commerce for Oceans and Atmosphere (Retired)

Wednesday, September 28, 1988

4.0 Concurrent Workshops
    4.1 Wind, Wave and Swell (Conference Rm 1C)   Leader: Jon F. Klein, Sea-Land Iselin, New Jersey

    4.2 Tropical, Extra-tropical Storms (Conference Rm 2A)   Leader: Kenneth A. Blenkarn, Consultant, Tulsa, Oklahoma

    4.3 Ice Currents, and Ocean Processes (Conference Rm 5A)   Leader: Allan R. Robinson, Professor, Harvard University, Cambridge, Massachusetts

    4.4 Nearshore Forecasting Conference Rm 5D)   Leader: William G. Gordon, New Jersey Marine Sciences Consortium

    4.5 Collection, Reporting Dissemination and Display   Leader: Richard Wagoner, National Weather Service

BREAK

Concurrent Workshops Reconvene

Lunch (Buffet) - Refectory

Concurrent Workshops Reconvene

BREAK

Concurrent Workshops Reconvene

ADJOURN

Thursday, September 29, 1988

5.0 Reports of Working Groups
    (Auditorium)

5.1	Wind, Wave, Swell	Jon F. Klein
5.2	Tropical, Extratropical Storms	Kenneth A. Blenkarn
5.3	Ice, Currents, and Ocean Processes	Allan R. Robinson
5.4	Near-Shore Forecasting	William G. Gordon
5.5	Collection, Dissemination and Display	Richard Wagoner

BREAK

Working Group Reports Reconvene in Auditorium

6.0 Discussion of Reports

ADJOURN NATIONAL MEETING

Appendix E:
Working Group 1: Wind, Wave, and Swell

JON F. KLEIN, Sea-Land Service, Inc., *Leader*
VINCENT J. CARDONE, Oceanweather Inc.
FORREST A. MILLER, Inter-American Tropical Tuna Commission
D. B. RAO, National Meteorological Center, National Oceanic
   Atmospheric Administration
ALLAN M. REECE, Shell Development Company
WILLIAM S. RICHARDSON, National Ocean Service, National Oceanic
   Atmospheric Administration

SOURCES OF FORECAST INFORMATION AND DATA

The group identified four major sources of wind, wave, and swell data: U.S. Air Force, U.S. Navy, NOAA, and all foreign centers (taken collectively). The Air Force was not given detailed consideration since their global forecast center operates atmospheric models only. Additionally, the Air Force does not generally release data to the public sector. Although foreign sources may play a greater role in forecasting in the future as exchanges of forecast products increase in number, this interaction is presently limited in scope. Therefore, detailed consideration of foreign source data was not pursued by the working group.

The U.S. Navy Fleet Numerical Oceanography Center (FNOC) provides the Navy with a comprehensive array of marine forecasts. Most of this information is directed toward satisfaction of the Navy's internal operating requirements. Selected FNOC products, including global and regional wind and wave analysis and forecasts, are distributed in real-time transmissions to the civilian sector through CNODDS and Radiofax. The information released by FNOC consists of graphic and tabular output from the Navy's numerical atmospheric weather prediction (NWP) models, its global spectral ocean wave model (GSOWM), and its regional Mediterranean spectral ocean wave model (MSOWM). These data require further analysis and interpretation before they can be productively used in site-specific forecast applications.

The Navy derives both analyzed and forecast surface-wind data from sophisticated planetary boundary layer formulations in its global (Navy Operational Global Atmospheric Prediction System—NOGAPS) and regional (Navy Operational Regional Atmospheric Prediction System—NOAAPS) prediction systems. This surface-wind information is disseminated to both military and civilian users through a variety of communications links. These wind data are also used to drive the Navy's global and regional wave models.

The Navy's global wave model (GSOWM) is a deep-water model that runs on a 2.5° latitude and longitude grid and represents the wave spectrum in 15 frequency and 24 direction bands. The GSOWM produces two forecasts per day extending out 48 and 72 hours. Many output products including sea and swell height, period, and direction are derived from GSOWM spectra and disseminated to various users.

MSOWM is a first-generation deep-water model run by the Navy covering the Mediterranean area in 70 km grids. The model represents the spectrum in 15 frequency and 12 direction bands. MSOWM also produces two forecasts per day and its spectral data receives dissemination similar to GSOWM.

NOAA also operates a global, deep-water ocean wave prediction model. Both the Navy and NOAA models are driven by forecast wind fields derived from different atmospheric models. The wave models themselves also differ in certain aspects regarding the treatment of the underlying physics. While the Navy's models are designed and operated primarily to satisfy the Navy's internal requirements, NOAA's wave modeling program is an integral part of NOAA's broader ocean products center, which is charged with serving the civilian community. Both the NOAA and Navy models undergo periodic monitoring, validation, refinement, and extension to provide higher resolution and better physical representations. For example, NOAA has implemented and the Navy is now implementing new wave models that include new physics and shallow-water effects.

NOAA produces ocean surface (10 m level) wind forecasts by applying diagnostic boundary layer procedures to the large-scale wind forecast produced by NMC's global atmospheric model. These wind forecasts are provided to National Weather Service (NWS) Forecast Offices (WSFO) using AFOS, AKFAX, and DIFAX. In addition, these forecasts are available on NOAA's Family of Services to all civilian users.

The NOAA operational global model is a deep-water model that predicts the two-dimensional wave spectrum on a 2.5° latitude and longitude grid. The model uses parameterized wave-wave interactions and has 15 frequency and 24 directional bands. The forecasts are issued once a day and extend for 72 hours. Output products are in terms of significant wave heights, primary period, and direction and are disseminated to WSFO's and civilian users via the same media mentioned above for wind forecasts. Two-dimensional spectra messages are also provided by NOAA for selected points along the U.S. coast.

As mentioned previously, NOAA has implemented its shallow-water spectral wave model to forecast waves over the Gulf of Mexico. In contrast to the global ocean wave model, this model takes into account the effects of bottom topography and bottom friction on wave propagation and dissipation. This model uses a 50 km (approximate) grid and is run twice daily to produce forecasts out to 48 hours. The output products, including significant wave heights, primary period, and direction are disseminated via AFOS and FOS. Plans are under way within NOAA to extend this shallow wave model to other U.S. coastal regions during the next 12 months.

In addition to dynamically generated wind and wave forecasts, NOAA also produces statistically driven wind and wave forecasts. A model output statistics (MOS) technique provides wind forecasts out to 48 hours at 91 coastal stations and more than 12 Great Lakes regions. Wave forecasts, based essentially on a SMB technique, are provided for 64 points on the Great Lakes and 6 points on the chesapeake Bay.

A summary of the sources of numerical forecast guidance of wind, waves, and swell is provide in Tables E-1 and E-2 which follow.

PERCEIVED USER REQUIREMENTS

The working group next attempted to delineate forecast requirements for various user groups. These user groups were loosely structured under four areas: coastal, nearshore, high seas, and the military. Requirements were then developed for each user area covering forecast and data parameters, spatial resolution, temporal resolution, forecast horizon, and delivery medium.

In general the working group found that each user group had significantly different forecast and data requirements. Each user group's requirements are equally critical to that group's needs. As noted in Table E-3, current marine weather forecasts do not meet these perceived requirements.

Coastal users, including recreational boaters and fishermen, commercial fishermen, marinas, port authorities, and others, require highly accurate

TABLE E-1 Sources of Numerical Forecast Guidance of Wind, Waves, and Swell

U.S. Air Force	Foreign Sources	U.S. Navy
Numerical weather prediction Global weather center • atmospheric only • no release to public	WMO Publication Number 12 Available on Radiofax Weather & wave forecasts Ad hoc combinations	NWP at FNOC GSOWM (Global Spectral Ocean Wave Model) • 2.5 x 2.5 degree grid MSOWM (Mediterranean Spectral Ocean Wave Model) • 70 km grid encompassing the Mediterranean • Models run twice per day with 72 hour range OUTPUT PRODUCTS: • Significant wave heights • Primary period & direction • Secondary period & direction • Sea & swell period & direction • Two-dimensional spectra • White cap index • Surface wind speed & direction DISSEMINATION: • CNODDS (Civilian, Navy, NOAA Data Distribution System) • Radiofax • Internal DOD

forecasts covering tight site-specific grids with a forecast horizon of 0 to 12 and 24 hours.

Conversely high-seas users, which include ocean shipping, ocean mining, commercial fishing, ocean towing, and other groups, require forecasts with a 20 percent accuracy factor, covering a grid of approximately 150 nautical miles and extending from 24 to 72 hours out to 6 to 10 days.

The nearshore users, which include the oil and gas industry, coastal shipping, waste disposal, commercial fishing, and other groups, have perceived requirements falling in-between the coastal and high-seas groups. The military's requirements span all three geographic areas and include other special mission oriented requirements.

The working group concluded that none of the perceived requirements for coastal users were being met, although a significant value-added effort using existing data could result in a marked improvement in coastal forecasts. Additionally, NEXRAD (NOAA's coastal radar program) will provide valuable data for this area, although it will also require expert interpretation by government or private sector value-added groups before it is useful to civilian end users.

The expansion of the GMWAVE model to other coastal ranges during the next year should allow the 50 km spatial requirement of nearshore users to be met, although two-dimensional information will still not be readily available to civilian users. The accuracy requirements of the nearshore user group is currently not being met.

With regard to high-seas users, the longer range forecast horizon, which is critical to effective voyage planning, is not being met. There is no formal program to address episodic waves or to implement the perceived need for a real-time warning program for episodic waves.

Although covered in detail by another working group, it was felt that certain comments concerning dissemination were appropriate with regard to wind, wave, and swell forecasts. The media through which wind and sea-state analyses flow to the end user fall into three classes: direct data links, NWS, and the private sector.

Direct Data Links

These links make possible the acquisition of vast amounts of observations and Navy and NOAA analysis and forecast products by the end user. The major limitation of this medium is the restricted or limited usefulness of raw model output for most forecast applications and the limited access to these links from offshore locations. Additionally, in 1985 so many users were accessing the Navy's NODDS data link that the system was failing. To protect the system, civilian access was discontinued. In an effort to maintain civilian access to this data, NOAA established the CNODDS program

TABLE E-2 NOAA Sources of Numerical Forecast Guidance of Wind, Waves, and Swell

NWP at Suitland	GMWAVE	MOS
NOW (NOAA Ocean Wave Model) • Global Deep Water Model • 2.5 x 2.5 degree grid • Run once per day • 72-hour forecast range OUTPUT PRODUCTS • Significant wave heights • Primary period & direction • Secondary period & direction • Two-dimensional spectra (at 18 coastal points) • Surface wind & direction • FOS DISSEMINATION • AFOS • AKFAX & DIFAX • FOS (Family Of Services)	Gulf of Mexico Model • Shallow-water effects • 50 km x 50 km grid • Run twice per day • Scheduled for expansion to east, west, and Alaskan coasts OUTPUT PRODUCTS • Significant height • Primary period & direction DISSEMINATION • AFOS	Statistical Wind Forecasts • 91 coastal locations • 12 Great Lakes locations • 48-hour forecast range EMPIRICAL WIND & WAVE FORECAST • 64 Great Lakes points • 6 Chesapeake points SATELLITE DERIVED LOW CLOUD TRACK • 50 South to 30 North • Every six hours • For analysis only

without additional funding. NOAA's long-term goals are to make all non-classified Navy products available through its Family Of Services program and discontinue its CNODDS program.

National Weather Service

The National Weather Service is the major source of user friendly forecast information. NWS offices provide coastal forecasts, offshore forecasts, and high-seas forecasts as well as gale and storm and special marine warnings. Under NOAA, the NWS attempts to meet user requirements within tight budgetary constraints and physical limitations of modeling data and other resources. In addition to the above example, the Navy has announced its intention to discontinue its East Coast facsimile (FAX) transmissions to the civilian sector. These are the primary source of forecast data for many commercial fishermen and nearby coastal shipping. NOAA must now step in and assume these transmissions, without additional funding, or this valuable information will be lost.

Private Sector

One area of the private sector, also known as "data brokers," typically operate as redistribution centers for graphical, satellite, and radar imagery, as well as alphanumeric data. This information is mainly acquired from NOAA's Family of Services facility at Suitland, Maryland, and via NEDIS (direct satellite data). The information provided by data brokers is modified in format but not content.

The value-added component of the private sector is a more varied mix of companies. Some provide very specific local forecasts, such as the dial-up surf forecasts available in southern California. Others provide site-specific wind and wave forecasts to contracted offshore oil and gas industry facilities, while a small number of companies provide high-seas forecasts to contracted companies in the form of voyage routing services.

A very small sector of the value-added group have demonstrated dedicated area and application specific computer-assisted wind and wave forecast systems incorporating human input for interpretation, quality control, and accuracy enhancement.

Other very small segments have recently demonstrated "expert" type systems that utilize on-board microprocessors and incorporate forecast wind and wave information along with vessel stability criteria to assist in decision making with regard to optimum routing, seakeeping, and damage avoidance.

TABLE E-3 Perceived User Requirements

User Groups by Area	Parameters	Temporal Resolution	Temporal Resolution	Forecast Horizon	Accuracy
COASTAL (0-50 Mi) • Recreational fishing • Recreational boating • Commercial fishing • Dredging • Port authorities • Waste disposal • Marinas • Oil & gas	Wind speed & direction Significant height Spectral peak period Superstructure icing	10 nmi[a]	3-6 hr	12-24 hr	very accurate at lower limits Critical threshold o 15-25 kn winds o 6 ft wave height
NEARSHORE (50-100 Mi) • Oil & gas • Commercial fishing • Coastal shipping • Waste disposal	Wind speed & direction Significant height Spectral peak period Two-dimensional spectra Superstructure icing	25-50 nmi	6-12 hr	24-72 hr	20% standard error
HIGH SEAS • Ocean shipping • Ocean mining • Ocean towing • Oil & gas	Wind speed & direction Significant height Spectral peak period Two-dimensional spectra Episodic waves	150 nmi	12 hr	24-72 hr, plus 6-10 days	20% standard error
THE MILITARY	All above parameters Sea/swell height Sea/swell direction Sea/swell period Ambient noise	75 nmi--global 30 nmi--regional	12 hr--global 6 hr--regional	120 hr--global 72 hr--regional	Various

[a] nmi = nautical miles

CONCLUDING REMARKS ON BENEFITS

Members of each identified user group feel meeting their specific forecast requirements is critically important for reasons of safety and economics. In the highly competitive high-seas market, a high benefit-to-cost ratio can be expected due to the level of potential savings in industries with very high fixed costs. This ratio is offset somewhat by the fact that the high-seas region has a relatively small population.

Solutions to meet the perceived requirements of the nearshore users group will also have a relatively high unit cost. The nearshore constituency is considerably larger than the high-seas constituency, and there are also significant benefit factors to be considered relative to the impact of improved forecast on environmental issues within the oil, gas, and waste disposal industries.

It was the working group's conclusion that the coastal region would likely derive the greatest benefit of meeting perceived forecast requirements. This benefit is measured in terms of protecting the lives and property of an ever-growing group of small boat operators, fishermen, and individuals engaged in other activities in this heavily populated constituency. Although meeting perceived requirements in this region will result in a low unit cost benefit, the total bottom-line benefit is greatest due to the overwhelming size of the coastal constituency.

ATTENDEE COMMENTS

The working group presented its preliminary findings on the last day of the workshop in Irvine, California, on September 29, 1988. Following this presentation, several attendees submitted written comments concerning the forecasting of wind, waves, and swell. The group did not meet again following its presentation and could not fully incorporate the submitted comments into its report. The author has, therefore, highlighted these comments to ensure that these comments are taken into account by the committee in preparing its final report.

A fishing industry representative (Atlantic and Gulf areas) stressed heavy dependence on FAX transmissions of wave heights and other data used in performing analysis at sea. There was a great deal of concern that regularly scheduled FAX transmissions not be discontinued.

A routing service manager from the value-added sector addressed the hazard of superstructure icing and its related forecast requirements. Superstructure icing is an extreme hazard when it occurs. It is presently forecast by means of a nomogram with entries including estimated air temperature, surface wind speed, sea temperature, and the duration of the event. The output is a forecast of anticipated superstructure icing in centimeters. The manager commented that the nomogram highly depends on the output of surface forecasts (except sea temperature) and could benefit greatly from improved forecast data.

A shipping company executive addressed the perceived need for disseminating more information regarding episodic waves and recommended implementation of some form of a real-time warning system for these dangerous wave systems.

An oil industry representative offered the opinion that "with the exception of the observation of sea ice, remote sensing technology is not in the position of making much additional contribution to the ocean data base in a reasonably near-term time frame." He went on to point out that experimental programs in this area are valuable and deserving of consideration for support, but should not receive an inappropriate amount of emphasis.

A U.S. Coast Guard (USCG) representative added the following comments concerning perceived needs applicable to USCG operation.

- Highly accurate forecast information including wind drift is required for the coastal and nearshore areas to support oil and waste spill containment and cleanup operations.
- This same highly accurate forecast information is also required to support search and rescue missions with specific needs in the area of wind drift predictions of drifting boats and objects and to support the decision-making process as to what equipment (e.g., boat size, helicopters) to allocate to rescue missions.
- Highly accurate wind and sea-state forecasts are required to properly plan buoy placement and tending operations.
- Accurate wind drift forecasts are needed to support ice operations in both the nearshore and high-seas areas.
- Regarding the USCG's military mission it was pointed out that accurate wind and wave forecasts are required to plan and execute port approach operations, that is, minesweeping.
- Spatial resolution, temporal resolution, forecast horizon, and accuracy requirements were detailed in the USCG's comments and covered all ranges of the perceived requirements outlined Table E-2 (depending on the area and specific mission application).

An unidentified attendee remarked on his perceived need for a strategically located network of fixed platforms and data buoys within 300 miles of the coast. He felt this network would significantly improve wind data gathering and wave prediction and modeling. He felt this network could fill the need in this area until an operational all-weather satellite was available to provide this data, and after such a satellite became operational the network could serve as a backup to the satellite and assist in calibration operations.

Appendix F
Working Group 2:
Tropical and Extratropical Storms

KENNETH A. BLENKARN, Consultant, Leader
ROBERT T. BUSH, Universe Tankships, Inc.
PAUL JACOBS, Office of Meteorology, National Weather Service,
  National Oceanic and Atmospheric Administration
CHESTER JELESNIANSKI, National Weather Service, National Oceanic
  and Atmospheric Administration
SAUNDERS A. JONES, Puerto Rico Marine Management, Inc.
POLLY MIRKIVICH, The Texas Shrimp Association
DAVID J. H. PETERS, Conoco, Inc.
FREDERICK H. SHARROCKS JR., Federal Emergency Management
  Agency

This report examines forecasting issues related to tropical and extratropical storms. The two different kinds of storms were examined as separate issues by the working group for several reasons. Tropical storms (hurricanes) and extratropical storms involve different meteorological phenomena. The forecasting organization and intensity of focus on marine aspects are different for hurricanes. The two kinds of storms affect different user groups or affect the user groups differently.

There is some overlap of concern with other working groups, particularly the working group on wind, wave, and swell and the working group on collection, reporting, dissemination, and display. To the extent possible, this report focuses only on storm-related user needs and responses. A major concern of the working group is with the impact of storm forecasting on safety.

FINDINGS

1. Forecasting by the National Weather Service (NWS) and user response have been successful in minimizing loss of life due to hurricane occurrence in U.S. coastal regions.
2. User groups are aware of the uncertainties of hurricane forecasting and generally accept the burdens of false alarm evacuation for hurricanes.
3. There seem to be low prospects for technical improvements in forecasting hurricane tracks and behavior until all-weather satellite remote sensing is developed and made operational.
4. Decision makers of user groups and local authorities do work well to manage evacuations to ensure safety yet minimize economic loss.
5. For the foreseeable future, opportunities for making improvements and maintaining public confidence lie mainly in planning and management of hurricane evacuation.
6. Rapid dissemination of real-time hurricane data from satellites, reconnaissance aircraft, and data buoys is vital to near-term forecasts and evacuation decision making for many user groups.
7. Significant improvements in accuracy and reliability of forecasting for extratropical storms would improve the operation of the fishing and shipping fleets.
8. The potential for improvements not withstanding, present marine forecasting of extratropical storms by the NWS is considered satisfactory by the fishing and shipping fleets.
9. The primary threat to shipping, and to many fishing vessels, is from "surprises," that is, from events that are qualitatively different from expectations based on forecasts.
10. To avoid surprises the masters of both fishing and shipping vessels depend very strongly on their own experience and judgment and on real-time weather data on board or from ships in the region.
11. There is a particular need to collect and disseminate weather observations from ships in storms and to do so at more frequent intervals than normal ship observations.
12. Marine weather conditions involving either explosive cyclogenesis or episodic waves may play a significant role in extratropical storm surprises.
13. While satisfied with marine forecasts, the fishing fleets do identify shortcomings in the NWS forecast product.
14. For the foreseeable future offshore petroleum industry exposure to extratropical storms is basically limited to Gulf of Mexico winter storms.

15. In spite of broad success in avoiding losses there remains some offshore industry vulnerability to unanticipated extratropical storm events.

16. Forecasting for future petroleum operations in remote, arctic offshore areas may be very dependent on weather data from improved satellite systems.

RECOMMENDATIONS

1. Federal agencies and local authorities or users should place emphasis on improving evacuation planning and decision making.

2. To support safe and effective evacuation decision making the NWS needs to ensure that real-time, or near-real-time, hurricane weather data gathering and distribution are maintained or even improved, such as by installation of improved coastal radar.

3. Some priority should be given to local measurement of hurricane storm tide measurements to verify and improve simulations and predictions.

4. As technology for satellite remote sensing of marine weather is developed, that technology should be rapidly mobilized to provide operational capability and make ocean weather data available to NWS forecasters and other marine users.

5. The NWS should revitalize and reinforce the ship observation system in order to

- obtain 3-hour or hourly interval observations from ships in storm areas;
- improve the quality of observations and reduce the time lag between observation and the reception by forecasters, specifically promoting the use of automated equipment such as SEAS units; and
- evolve procedures and communication systems necessary to provide storm weather data rapidly to vessel masters in the area of a storm.

6. To improve effectiveness of forecasts and of user operations, the NWS and cooperating agencies should increase interaction and communications with user groups; for example, strengthening activity of port meteorology officers and contact with fishing fleet groups.

7. While continuing efforts to advance basic technology regarding explosive cyclogenesis and episodic waves, the NWS should exploit emerging opportunities to provide mariners with guidance and warnings of such phenomena.

8. The NWS and offshore petroleum operating companies should reach accommodation and understanding to provide more offshore weather measurements and to utilize advanced radar systems.

9. Agencies concerned with ocean weather should reach agreements to establish a dedicated function responsible for improving the voluntary observing ship program.

TROPICAL STORMS

Forecasts

Within the NWS there is a special organization, the National Hurricane Center in Miami, dedicated to the development of hurricane forecasts and the associated warnings to the populations and forecast users affected. The geographical regions of responsibility include the eastern North Pacific, but the bulk of activity is concerned with the western North Atlantic and adjacent seas.

Over the years the NWS has developed computer simulations of hurricane processes and special hurricane data gathering methodologies. In performing the computer simulations for forecast purposes the National Hurricane Center utilizes weather data from the normal worldwide weather data network. In addition, special attention is devoted to interpretation of satellite sensing and imagery data, and special reconnaissance aircraft are employed to gain direct atmospheric measurements within the hurricane itself.

The center combines simulations and data to develop present condition descriptions of a hurricane and to project future changes in location, size, and intensity. Special priority is placed on identifying the sections of coastline expected to be influenced by landfall of the hurricane, the wind and tide to be experienced during passage of the hurricane, and the timing of such conditions.

Shoreside Response

The primary concern of shoreside populations and industry is to guard against loss of life during a hurricane. Consequently, shoreside communities in the coastal zone of expected landfall are generally "boarded up" and evacuated well in advance of a hurricane arrival. Residents of coastal areas subject to hurricanes have come to recognize and accommodate the uncertainties in the Hurricane Center forecasts and advisories. There have been some exceptions, but the overall history within the United States has been one of remarkable success in minimizing loss of life in hurricanes. Most exceptions have been attributable to individual exercise of poor judgment (defined by a consensus of what experienced, long-time residents would deem to be poor judgment). It must be noted that within the United States, government authorities are not empowered to forcibly remove private citizens.

The administrative processes for issuing evacuation advisories and the particular public authorities who may be responsible may vary significantly from one coastal region or community to another. The levels of planning and technical knowledge about local conditions may also vary widely. The Federal Emergency Management Agency (FEMA) has supported many local communities in developing guides to evacuation decision making and plans for evacuation.

Such studies have included simulations by the NWS, with a computer program called SLOSH, to describe the details of anticipated local storm tide flooding for various levels of hurricane intensity and paths of approach.

Storm flooding is a key measure of hurricane damage potential, threat to life, and urgency of evacuations. Many coastal communities have not yet undertaken such formalized evacuation planning studies.

Shoreside response also includes the securing and shutting down of various industrial operations such as harbor facilities, power stations, chemical plants or refineries, and petroleum production. For many of these operations, shutdown and restart are very costly. In such cases, evacuation often involves a very carefully staged process of gradually securing facilities and reducing staff well ahead of a hurricane approach, with final shutdown and evacuation of the last few key personnel being delayed as long as prudence allows. The object of such intense planning and management is to minimize the economic loss attributable to fake alarm shutdown. Many industrial operations may employ value-added forecasting efforts to fine tune decision making and evacuation scheduling. Value-added efforts are particularly employed for offshore oil and gas operations.

Fishing Fleet Response

Fishing fleets are very attuned to and respectful of hurricane warnings and advisories. Fishermen operate with the objective of being in port, secured, and evacuated before arrival of a hurricane. The consensus attitude of the fishing fleets is that the Hurricane Center provides appropriate and satisfactory forecasts and the history of the fishing fleets has been one of success in avoiding loss of life.

Recent success comes in the face of some opposing influences. Quite apart from the threats of adverse weather, fishing tends to be an uncertain and risky business. There are strong economic pressures for fishermen to try to minimize the unproductive time spent in port without income. Even worse, sometimes there is a special attraction to fishing in advance of a hurricane. Fishermen have observed that changing ocean conditions at such times changes the movement patterns or behavior of some target species to permit especially profitable catches. Thus some fishermen may be tempted to delay seeking shelter until the last moment. In a qualitative sense, the fishing fleet experiences the same incentive as industrial establishments to maintain operation as long as possible. The difference is that there is more safety risk for fishermen, and they are not organized to have value-added and focused forecasting services. One must not overemphasize the hazards, but the U.S. Coast Guard occasionally rescues fishermen in danger from a hurricane.

Shipping Response

Commercial merchant cargo and bulk vessels are generally very careful to monitor and respond prudently to hurricane forecasts. It is common practice to alter course or speed and delay sailings to give hurricanes a wide berth. This practice appears to be very successful in avoiding threatening conditions in the Caribbean and Gulf of Mexico as well as along the Atlantic Coast. In these areas hurricanes are intensively studied and tracked by the National Hurricane Center and response options for ships are numerous. In the mid-Atlantic, response options may be reduced and forecasting attention may be less, so that the overall experience, while basically successful, is not so comforting. The sinking of the *Derbyshire* in the western Pacific illustrates the practical problems of avoiding tropical storms in the open ocean. Special mention is made of the difficulties in obtaining adequate forecasts in such areas as the Indian Ocean, outside of the regions served by the NWS forecast system. This reinforces the consensus that in forecasting of tropical storms the NWS effectively performs its mission in support of ocean shipping through its domain of responsibility.

Offshore Petroleum Response

Since the inception of activity in the Gulf of Mexico, the offshore petroleum industry has been very cautious in trying to avoid risks to life posed by a hurricane. The accepted practice is to evacuate most personnel from offshore locations well in advance of a hurricane. Industry evacuation can involve a fleet of several hundred helicopters, supplemented by workboats and crewboats, to move the offshore work force of several thousands onto shore in time for evacuation in an orderly way along with a substantial complement of shoreside support personnel. A common yardstick has this process under way 72 hours ahead of the hurricane.

The offshore industry has become increasingly confident in the National Hurricane Center guidance and has gradually become somewhat more selective in the operating areas to be shut down and evacuated. Special attention is paid to the Hurricane Center forecast of storm track and

TABLE F-1 Representative Economic Loss of an Offshore Hurricane Evacuation

Action	Cost
Daily oil production shut-in	800,000 bbl per day
Representative shut-in term	3 days
Price of oil shut-in	$15 per bbl
Representative value of oil shut-in[a]	$36 million
Representative present worth loss[b]	$18 million
Helicopter rental and operation for evacuation	$5 million
Mobile offshore drilling unit rentals during shutdown (unproductive expenditure)	$5 million

[a] Treated as "loss" of product.
[b] Treated as deferral of revenue stream to depletion, 10 percent annual production decline, 10 percent discount interest rate.

projected zone of expected landfall. The direct economic costs of an evacuation are substantial. Table F-1 illustrates this with an estimate of the costs incurred for evacuation recently for hurricane Gilbert. Moreover, we must recognize that the logistic operation of quickly moving large numbers of people from offshore locations is not itself a totally risk-free undertaking. There are strong reasons to avoid unnecessary evacuations.

The industry remains very cautious on safety, but steps are taken especially to reduce the economic impact of shut-in of productions. Moves in the direction of more automation and remotely signaled closing of wells and treating facilities have permitted production to be maintained at least until the onset of hurricane conditions. Gas wells in particular often remain in operation throughout an evacuation. In addition, some petroleum operating companies employ intensive value-added forecasting efforts in order to support evacuation decision making and the optimization of evacuation timing. During the opening day of the workshop, David Peters, staff meteorologist with Conoco, described the program of his organization to provide specifically focused and continuously updated forecast advice to operations decision makers. It must be emphasized that these value-added efforts supplement rather than duplicate the information from the NWS. For example, a major part of the effort is concerned not with hurricane extreme conditions but on near-term local forecasts of conditions on the fringes of the storm under which evacuation operations are conducted. The extent of value-added service employed throughout the industry varies with the size and technical strength of the operating organization and the economic and safety consequences of evacuation.

The contributions that value-added services can make to safety assurance and reduction of economic loss critically depend on the hurricane meteorology data disseminated by the NWS. *Specifically, it is considered essential that the NWS continue with its present practice* of gathering, processing, and rapidly transmitting basic data from data buoys, observing ships, satellite imagery, and reconnaissance aircraft flights through the hurricane.

Needs and Priorities

The NWS forecasting of hurricanes and user response has to be judged a success in guarding public safety. During the past two decades the number of deaths atibutable to hurricanes in the U.S. coastal regions has been very small. This has been achieved by timely warnings and evacuations. Given the uncertainty in the forecasting of hurricane movement and behavior, the Hurricane Center warnings reflect a generous margin of error to ensure safety. One consequence is that a high percentage of the areas evacuated do not actually experience damaging hurricane conditions. These "false alarm" evacuations do represent financial loss to individuals and a burden to the local community generally. Nevertheless, most experienced residents of coastal communities view such losses as a necessary cost of living in hurricane-prone coastal areas.

Recent success in guarding against the dangers of hurricanes could be grounds for contentment. There are, however, some shortcomings in the overall national management of the hurricane threat.

- There have been some deaths blamed on fast moving or erratic hurricanes that have caught victims in exposed conditions.
- There have been incidents of multiple false alarm evacuations in rapid succession. Evacuation compliance has tended to diminish for the later storms, fortunately without bad consequences so far.
- There has been some public controversy over evacuation decision making; for example, regarding evacuation of Galveston during hurricane Gilbert. Such occasions can eventually undermine public confidence and proper response to warnings.
- There are several large population centers where the evacuation process has not passed a serious test.

It is important to maintain or even increase public confidence in hurricane evacuation management. Improvements in the accuracy of hurricane forecasts would naturally contribute markedly to this objective. Moreover, improvements in user community response would improve safety.

The critical uncertainty in the forecasting of hurricanes lies in the prediction of storm track. The past 20 years has seen a steady accumulation of hurricane data and research to improve mathematical simulation of hurricanes and general rules for prediction. These efforts do not seem to have brought any really significant improvements in the reliability of track predictions. The technical challenge is indeed formidable. There may be fundamental randomness (or natural chaos) that can never be avoided. However, at the current state of technology, it would appear that the dominant barrier to improvement is the paucity of input data available. This appears to be the case for general weather forecasting as well as for hurricanes. Hence it is not reasonable to expect that a greatly intensified technical effort in hurricane track research would bring commensurate progams.

The best prospects for improved hurricane forecasting may emerge from advances in satellite technology and operation that would permit all-weather remote sensing of meteorological data, especially over the oceans. Should satellite technology make adequate arrays of input data available, forecast accuracy can be expected to improve with present simulations. Even more improvements could then be expected from a major effort to develop more rigourous simulations with the expanded input data. However, for now, forecast improvements regarding hurricane track are on hold.

Low expectations for hurricane track forecast improvements in the near future tend to focus technology opportunities to enhancing user response. For many coastal areas, storm tide response depends on the local coastline, bay, or estuary configuration. Therefore the generalized hurricane tide forecasts by the NWS may realistically address the extent of flooding that should be expected locally. The NWS computer program SLOSH can simulate the local characteristics of storm tide. It is not, however, within the resources of NWS to provide real-time forecast simulations of storm tide at local sites during a hurricane approach. On the case studies performed with the SLOSH program, in collaboration with FEMA, the tide response has been simulated for representative hurricane conditions. Results permit local authorities to include realistic flooding scenarios in evacuation planning and to interpret Hurricane Center forecasts so as to include anticipated local storm tide in evacuation decision making.

Quite apart from political questions about jurisdiction and funding of planning studies, the working group concludes on pure technical and engineering management grounds that

- local evacuation planning studies are appropriate and are likely to increase safety and reduce economic loss; and
- for many local areas, storm tide simulations should be performed to support planning studies and decision making.

Whether storm tide simulations are made with the NWS SLOSH program or with a comparable analysis, it will probably be beneficial to make local storm-tide-related measurements during hurricanes. These measurements would permit verification of simulations, and, for example, provide assurance that particular locally important phenomena, such as wave setup, are reasonably reflected in the simulation. Verfication and local calibration may be important to ensure that erroneous tide simulations and predictions do not mislead decision makers or undermine public confidence.

With accumulation of experience in managing evacuation it is to be expected that decision makers, both public and private, will strive to optimize the process in terms of both safety and economic impact. More and more users are likely to utilize value-added special forecasting efforts of the sort described by David Peters of Conoco. Not only must NWS's basic data service continue to support near-term forecasts, but further technical advances should be pursued. Many of the near-term forecasts, important to evacuation, could be markedly improved with real-time data from Doppler radar. New radar (i.e., NEXRAD) should be deployed at appropriate coastal locations and NWS should collaborate closely with users to maximize benefits for evacuation decision makers.

EXTRATROPICAL STORMS

Marine forecasting of extratropical storms impacts mainly on the shipping industry and the fishing fleets. There is some influence also on offshore petroleum operations. Qualitatively, the needs of these user communities are similar, but there may be significant quantitative differences in the level of storm severity that is threatening. For example, winds and seas dangerous to inshore fishing vessels would represent only minor inconvenience to a tanker on the high seas. Hence the working group has viewed the distortions between storms and normal sea conditions not as a simple numerical threshold, but rather regarding the level of threat to the user.

Shipping

Maritime shipping has provided descriptions of ways in which improved marine forecasts would improve the overall effectiveness of shipping operations. Presentations on the first day of the workshop by Capt. Robert S. Murray of Matson Navigation and Capt. Saunders A. Jones of Puerto

Rico Marine summarized such benefits. Better intermediate forecasts (beyond 3 days) would make for better ship route planning. Improved shorter range forecasts would permit more effective operational decision making for course corrections to optimize an ocean passage by avoiding rough weather. More information and detail about storm characteristics would significantly improve management of the ship to reduce hazards or to minimize ship motions and cargo damage. The difficulties and risks of port approach and harbor passage would be reduced with more accurate and more frequent forecasts of local coastal conditions.

There is little doubt that significantly improved marine forecasts would have economic benefits for shipping and make ship operation easier. Indeed, one can reasonably speculate that very accurate and reliable forecasts might improve shipping in ways that cannot now be envisioned. The shipping community, however, considers the present forecast services as satisfactory and is hard pressed to identify quantifiable benefits of better forecasts or ways in which shipping operation would be changed in any fundamental way.

While expressing satisfaction with forecasts in the North Atlantic and North Pacific areas assigned to the NWS and cooperating agencies for support of international shipping, shippers note some difficulties with forecasts in other regions. In the Southern Hemisphere, especially on less heavily traveled routes, forecast services are often less than satisfactory. Particular mention is made to the difficulties in obtaining adequate forecasts in the Indian Ocean. There is some belief that the U.S. Navy as well as several other countries do provide marine forecast services superior to that received from U.S. civil sources. Canada, England, and Japan are almost universally noted as having superior services. Common observation is that Japanese broadcast fascimile maps can be received in a much clearer and more readable form. Fishermen as well as shippers note that Halifax marine forecasts are usually more accurate than adjacent U.S. services.

In contrast to the consensus that marine forecasts are generally satisfactory, one can emphasize that weather-related shipping losses continue to occur. Marine insurance organizations report losses of approximately 500,000 tons each year in incidents that are weather related.This constitutes approximately 30 percent of all shipping losses. Ship captains and shipping managements suggest that most of the weather-related losses can be attributed to the occurrence of "surprises" in weather developments. The word "surprises" is carefully chosen to convey that the issue is not one adequately captured in the usual assessments and measurements of the quantitative accuracy of forecasts. From the point of view of the mariner, a surprise event is a *qualitative, not just quantitative, variance* of the weather encountered from that anticipated on the basis of forecasts. Mariners consider the occurrence of surprises as the governing threat to safety of transocean passage.

Considered as a single class of events by mariners, surprises may in fact reflect several meteorological issues or phenomena. The working group tried to segregate the issues for assessment: open-ocean resolution, explosive cyclogenesis, and episodic waves.

Open-Ocean Resolution

There is a perception among mariners that many of the surprise events in the open ocean involve weather phenomena that are too small or too rapidly changing to be detected, processed, and reflected in forecasts of the NWS system. Based on this perception and much tradition, ship masters depend very heavily on their own judgment in formulating the forecast to specifically guide their decision making. Many of these masters have a practical working knowledge of meteorology phenomena and much experience in evaluating forecast projections for critical decisions. Attention on board ship narrows to the most important aspects of likely weather developments and tends to respond immediately to changes in local conditions. In addition, mariners express the need for more detailed information and insight regarding particular storms and especially to the need for more data on present weather conditions. This is not surprising, given the emphasis of ship captains on the need to do their own near-term forecasts.

It is almost unavoidable that meteorology specialists and forecasters within NWS must have some skepticism about the claimed merits of forecasting by individual ship captains. This would all seem puzzling since any forecasts made on board would lack any significant amount of computational power. Nevertheless, the preponderance of knowledgeable maritime opinion is that experienced captains do bring important forecasting capabilities into their decisions. The main problem is that these capabilities are employed at varying skill levels. Moreover, background in meteorology now seems to be less strong in the training of younger officers. This would suggest that in time the reliance on and effectiveness of shipboard judgment may decrease.

To provide extra support to ship captains, some shipping companies employ private, value-added, forecasting services. For the most part these value-added consultants supplement the basic NWS maps and forecasts. There are some differences of viewpoint in the meteorology community regarding the degree to which consultants can really bring enhanced skill levels to value-added forecasts. At the very least, a consultant can bring focused attention on weather along a specific client vessel's course and on that vessel's decision-making parameters. The consultant can give extra attention to assessment of all data reported in the region of interest. Opinions seem to be mixed in the shipping community as to the real usefulness of the value-added forecasters. Continuing and direct conversation between forecast consultant and the operations decison maker, a common practice in the offshore industry, does not appear to be common in shipping.

Quite apart from questions of shipboard skill or value-added enhancement, safety of ships would be increased through general improvement of open-ocean forecasts to minimize occurrence of surprise storm conditions. There are technology elements that have great promise for improving forecast effectiveness.

- Significant increases in ocean weather data density are technically feasible.
- Finer grid weather simulations would be possible with increased computing power.
- Advanced communication networks would permit rapid transmission of data and forecasts directly between ships and forecast centers.

Eventually, implementation of such technologies will come, not primarily for just marine forecasting, but as part of a overall global progress in weather data gathering and forecasting. Certainly gathering data over the ocean is a major part of data improvements generally, and satellite all-weather sensing may be the primary enabling technology.

The working group has tried to focus on near-term needs and options for improving marine forecasting to reduce the threat of weather surprises for shipping. The most fundamental need is for more ocean weather observations.

- NWS weather simulation and forecasts are particularly degraded by a lack of open-ocean weather data, especially in storms.
- Mariners place very high value on real-time nearby observations of storm conditions.

Open-ocean weather observations of storm conditions often critically depend on Voluntary Observing Ships (VOS). It is a major finding of the workshop that the overall program of VOS does not function as effectively as it should. This is particularly true of VOS data from storm regions. Therefore the work group on tropical and extratropical storms strongly recommends

> *that the Voluntary Observing Ship program be reinforced and augmented to make real-time ocean storm data rapidly available to NWS and other marine users.*

Working group deliberations identified several issues that would influence the effectiveness of an upgraded VOS program. Ship observations under storm conditions should be submitted at more frequent intervals than routine ship observations; for example, submit storm observations every 3 hours or at hourly intervals. NWS would need to define very carefully the threshold to qualify for storm conditions. Storm observations should carry a special designation to distinguish them from more routine observations. The NWS communication system and procedure must be modified to respond more quickly to process storm observations in order to

• permit NWS forecasters and value-added forecasters to follow storm conditions closely and issue warnings of changes; and
• quickly transmit storm observations data back to the shipping fleet for timely on-board decision making.

The present functioning of the VOS program is plagued by numerous problems of quality control, timeliness of reporting and communications processing, and shipboard procedures. For example, it seems that many ships report observations only during daylight watches because transmitting at night requires special overtime pay for the radio operator. The working group observes that impediments to weather observations for storm data should be removed in the interest of safety. These and other problems would be overcome by better mobilization of available measurement and communications technology. There are, of course, choices in equipment. As an effective first step, much progress would be made by placement of the NOAA-developed SEAS (Shipboard Environmental Data Acquisition System) units on the majority of observing vessels. A companion step would be placement of updated receiving units on ships, for example, to receive fascimile information, forecasts, and data by satellite.

Achievements of the equipment update objectives will call for a strong initiative from management of NWS and cooperating agencies to bring a renewed dedication on the part of shipping companies and observing ships. NWS should also strengthen the technical advice and support provided to observing ships regarding communications equipment. As a part of this strengthening support, NWS should enhance the interaction with ships at the technical and operating staff levels. *It is very important that port meteorological officer functions should be given higher priority for contact with ship personnel to enhance participation in the VOS program and improve effectiveness of communications to and from ships.*

Obtaining data from many scattered, even remote, operating locations is a challenge in many technologies and organizational settings, both public and private. An all too common approach is to specify the data that are desired, request that operating or line organizations provide the data, and then wait for data to flow in. The almost universal experience is that passive waiting will not maintain the required flow of data. It takes repeated or continual attention from the data receiving organization to ensure that data gathering and transmittal are adequately prosecuted. Such attention must

- verify that operating personnel can, in fact, expeditiously and easily gather and handle the data appropriately;
- establish a basis for adequate tutoring of operating personnel in skills required for data activity;
- perform at least some spot checking of data and calibration of equipment;
- provide feedback and encouragement to data gathering locations to maintain sufficient priority on data gathering; and
- maintain open and reliable communication channels for data.

Like the proverbial "free lunch," there is no such thing as free data. It requires nurturing attention from the receiving organization to keep data efforts functioning. Often the receiving organization only complains about the inadequacy of what is received. If the data are not sufficiently important for the receiving organization to contine required vigorous nurturing, the appropriateness of burdening operating personnel with data gathering is to be questioned.

The assessment of the disappointing effectiveness of the VOS program strongly suggests that the program suffers from a lack of nurturing. To achieve an effective VOS program, NOAA and the other agencies concerned with ocean weather should take initiatives to establish a dedicated function aimed at coordinating and nurturing the VOS program.

Improvements of effectiveness of the VOS program is the first priority in improving forecasting and response to ocean storms. There are still other steps for NWS to focus on the special needs of marine forecasting, such as

- increased forecaster training in the particular skills of ocean storm forecasting;
- increased forecaster awareness of the needs of marine users; and
- increased experience feedback from marine users.

Explosive Cyclogenesis

As a matter of quantitative meteorology, explosive cyclogenesis describes extraordinary, low-pressure systems that deepen at rates of 1 millibar per hour or faster. Such storms are not well forecast by NWS simulations of ocean weather. They occur in the Northwest and Northeast Atlantic as well as in some North Pacific regions. Evidence suggests that occurrence of explosive cyclogenesis is most common near coastlines. There are indications that the phenomenon involves physical processes not presently captured in atmospheric simulation, at least not in operational programs.

While the term "explosive cyclogenesis" is not used by mariners, there is reasonably wide recognition of the threat of fast developing events, termed "bombs" in marine user parlance. The occurrence of poorly forecast bombs is dangerous for shipping. Efforts to improve forecasting deserve high priority in the interest of safety of life at sea.

There have been special meteorology seminars on explosive cyclogenesis, and there is some research moving forward. This work should continue. As always in such matters, progress is hampered by a lack of data. Scarcity of data in the air column over the oceans seems to be a particular problem, and direct data from upper-air probes are specifically needed. It appears that this need can most effectively be addressed by ship-launched devices. To the working group this suggests that the Automated Shipboard Aerological Program (ASAP) should be prosecuted vigorously as a complement to efforts to improve effectiveness of VOS.

Without waiting for breakthroughs in the basic research, the NWS has developed some empirical guidelines to assist forecasters in recognizing and dealing with explosive cyclogenesis events. Parameter threshold values and indicative weather patterns are established to identify the likely occurrence of explosive cyclogenesis. Further evolution and application of these empirical methodologies should continue. At present it appears that the greatest need is for more aggressive coaching of NWS forecasters in using the established guidelines.

From the viewpoint of the mariner, it may be that NWS would improve service to users by specifically identifying storms with potential for explosive cyclogenesis. The simple statement, "This storm may be a bomb," would clearly alert ships to the particular uncertainties involved and to the need for extra caution. Increased education efforts are warranted to enhance mariner awareness and understanding regarding such events.

Episodic Waves

There is a body of anecdotal evidence to suggest the occurrence of waves dramatically larger than anticipated on the basis of the prevailing sea conditions. Technical specialists refer to such waves as episodic waves. The working group is aware of research that has been dedicated to such phenomena, in part sponsored by the U.S. Ship Structures Committee (with input from the Marine Board's Committee on Marine Structures) and in part by European institutions, especially in Norway, but there does not presently seem to be a clear resolution of questions about episodic waves. There does not for example appear to be technical consensus as to whether episodic waves reflect a particular physical phenomenon or are instead merely a manifestation of the statistical variability within a given sea state.

Irrespective of the technical debates, mariners view the occurrence of impressively large and unexpected waves as a real threat to safety at sea. The term "rogue wave" is commonly used to classify such events. The issue is not strictly a forecasting question. However, the practical consequences are sufficiently serious that research efforts in the broader physical oceanography and naval architecture community should continue. The NWS should remain aware of progress and seek insights and understanding. Any opportunities should be exploited to evolve toward useful warnings and to educate mariners about the phenomenon.

There are some areas of the world (i.e., off the coast of South Africa) where seas interact with strong currents to produce very steep and potentially damaging waves. This may or may not be considered an episodic wave issue. Nevertheless, mariners are keenly aware of the hazards of such steep waves and try to avoid conditions under which they occur. Some forecasting services do attempt to provide warnings of possibly dangerous wave-current interaction. The NWS should consider including this concern in the marine forecasts for the ocean areas for which it is responsible.

FISHING

The information examined by the working group concerned the Atlantic and Gulf fishing fleets. Subsequent inquiries suggest that the group's assessments generally apply to other U.S. fisheries as well.

In the North Pacific, Bering Sea, and Gulf of Alaska area, *Lloyd's List* reports more than 1,100 fishing vessels ranging up to 650-foot mother vessels with more than 200 people on board working in this area. Many of these vessels are home ported in Washington and Oregon. The replacement cost of a large modern processing mother ship can exceed $40 million.

Extended voyages of over a month's duration are common. About seven vessels are lost each year; however, during the period from November 1988 to January 1989, 15 vessels sank with many lives lost. This was partly due to exceptional weather with very strong winds and record low temperatures causing severe icing conditions. Icing of rigging and superstructure is a major cause of capsizing and foundering. Thus air and sea temperature and surface wind speed forecasts are critical during the winter to early spring period. Accurate forecasts of sea ice formation and movement are also very important to the fisherman to minimize losses of fixed gear, such as crab pots, and to indicate the limits of the ice edge. Fishing vessels are even more vulnerable than deep-sea ships to the exceptional waves that form in this area.

In general, inshore and offshore fishing fleets have different concerns. The inshore fleet is made up of small vessels, perhaps 25 to 40 feet in length, and these vessels are engaged mainly in day fishing trips. They are equipped with only minimum communications gear. Forecasts are generally received on very high frequency (VHF) voice broadcast or high frequency (HF) single sideband voice broadcast. Inshore fishermen also pay attention to local commercial television forecasters in decision making about plans for fishing.

The offshore fleet is composed of larger vessels, many longer than 100 feet, to perhaps 150 feet. The voyages of the offshore fleet may last as long as 4 weeks and cover 1,000 miles or more. These vessels monitor voice broadcasts for weather forecasts. However, the offshore fleets increasingly rely on facsimile broadcasts for forecasts.

For most fishing vessels, economic pressures push them to operate on the margins close to conditions threatening damage or safety. They are vulnerable to storm phenomena even smaller than trading vessels. Fishing vessel captains are very dependent on their own interpretation of local data. They tend to maintain ship-to-ship radio voice communications with other fishing vessels in the region of their location to exchange weather observations. Even with prudence and considerable skill the fishing business can be a nerve wracking struggle requiring acceptance of a significant amount of weather-related risk. There can be little doubt that major improvements in forecasts would increase productive fishing time and reduce risks for the fishing fleets.

The fishing fleets, in spite of difficulties and weather hazards, view the NWS in a very positive way and describe forecast services as satisfactory. Perhaps, contending directly with ocean weather daily as they do, fishermen are more aware of the inherent capriciousness of nature than are most landsmen. In any event, the fishermen look on the NWS as their protector, and feel they are well served. Although there is not very much direct contact between fishermen and NWS representatives, fishermen seem impressed with the helpfulness displayed by forecasters or other staff in the occasional direct request for assistance. Fishing fleet representatives do have some concern that the needs of fishermen may be overlooked as NWS changes policies, reorganizes the forecasting function, and modernizes communication equipment.

The general expression of satisfaction notwithstanding, fishing representatives have a menu of shortcomings of the forecast service that they believe deserve some attention.

- Fishermen find serious problems when they are presented with two very divergent forecasts, as sometimes happens when they are in the region of boundaries between two different NWS forecast areas.
- Like the shipping captains, fishermen would like more details about the nature of the storm system that they may encounter.

- Even recognizing the uncertainties, fishing captains yearn for more extended forecast information and more frequent updates of conditions during a storm.
- A system for rapid reporting and broadcast of current weather observations by the fishing fleets is needed.
- There is a need for more explanations when forecasts are changed drastically.
- Sea surface and water column temperature are especially of interest to fishermen.
- There is a recurring message: fishermen wish U.S. forecasts were as good as those from Halifax, Canada.

For many problems of fishing fleets, solutions may basically involve mobilizing available technology. This raises the question of the role of value-added services. Some fishing groups have engaged forecast services, but the outcome has most often not been satisfactory. The offshore fleet particularly may be scattered and far ranging, and it may be very difficult for a value-added consultant to target and communicate with such a fleet on an economical basis. The fishing fleets will continue to look mainly to the NWS for forecast needs.

It is not the mandate of NOAA and NWS to provide fishing fleets with the kind of specifically tailored guidance that a good value-added forecaster would provide. And it is not likely that public policy will change that mandate in the foreseeable future. While there is some contact betweeen fishermen and NWS, it is the assessment of the working group that a more concerted effort of contact and discourse bewéen NWS personnel and fishing fleets is needed. It is expected that there remain several potential adjustments of forecast operations and services that NWS could undertake to overcome shortcomings of forecasts for fishermen without departing from its mandate or from current NWS policy. Meetings with fishing fleet organizations could identify useful adjustments and could smooth user transitions as NWS makes organizational changes or modifies the forecast delivery communications system and enhances fishing fleet involvment in VOS.

For the inshore fleet, major improvement could come from progress on the general problem of local area forecast resolution. It is expected that available technology could bring such progress. The major barrier seems to be institutional and to a large degree a question of funding. It may be that progress on local forecast resolution will require local initiatives. To illustrate the potential role of local initiatives, several workshop attendees cited local surf forecasts available in southern California. Certainly the fishing fleet should be a part of any local forecast initiative.

OFFSHORE PETROLEUM

The offshore petroleum industry has conducted operations in most of the coastal regions of North America: the Gulf of Mexico, U.S. Atlantic Coast, Newfoundland-Labrador, the West Coast (especially southern California), Gulf of Alaska, Bering Sea, and Beaufort Sea. These operations have involved substantial vulnerability to ocean storms, especially exploratory drilling and construction. Because of this vulnerability, the offshore industry has extensively utilized value-added consultants to provide specialized forecasts to support operations. These forecasting programs have commonly had several particular attributes:

- frequent and continuing conversations between the forecaster and the onsite superintendent (decision maker) to ensure clear understanding of operating need and most useful interpretation of the forecast;
- continual feedback of onsite weather observations to the forecaster's office; and
- adjustment of forecast attention to emphasize the specific weather variable most important to operating decision making.

Industry forecast progress culminated in the forecasting effort to support deep-water exploratory drilling off the mid-Atlantic Coast. This program has been presented in the technical literature and was described in a workshop presentation by Allen M. Reece of Shell Oil Company. Very focused forecasts were targeted at predictions of specific drilling vessel motions. Onboard computer systems employed forecasts together with real-time data and simulations of ship motions to provide expert system guidance to storm preparation decisions. The overall forecast effort contributed to the prosecution of a very difficult drilling campaign without major storm mishaps. The offshore industry places high value on the potential benefits of value-added forecast services.

The foregoing is not to suggest that the offshore industry has been without storm-related accidents. The tragic sinking of the semisubmersible drilling unit *Ocean Ranger* on Grand Banks heads the list. There have also been several incidents in which jackup drilling barges capsized or collapsed during winter storms in the Gulf of Mexico.

Under current industry conditions, operations have been reduced to a significant degree. Lack of success in exploratory drilling has been the governing issue, not just the fall in oil prices. The result is that industry concern with extratropical storms has almost exclusively narrowed to the Gulf of Mexico winter storms. With the long experience history in the Gulf of Mexico, operating practices have been improved to reduce weather vulnerability, and there is less need for intensive value-added forecasting for many operations. In fact, many operating organizations now rely only on NWS forecasts.

Even with advances in operating practices, there remains a safety vulnerability for certain kinds of operations to frontal systems, local squalls, or other rapidly changing storm conditions. Jackup barges under tow, construction barges, workboats, and helicopters can be caught in exposed locations by severe weather and endangered.

Not suprisingly, the most immediate path to improving forecasts of storm events lies in expanding the presently utilized weather data base. The numerous offshore platforms in the Gulf of Mexico are at locations where weather observations would provide valuable additions to the NWS data base. Oil companies do make weather measurements on many platforms. Some of these data streams are sent to the NWS system, many are not. Operators of offshore platforms should do more to provide ocean weather observations, and there is a specific cooperative project under way to improve this response of the offshore industry. It should be noted, however, that there is another side to the story. The NWS has not always responded in ways that encourage petroleum operating companies to provide observations.

- Oil company staff and technical specialists who have been involved in offshore weather measurements often conclude that it is hard to give data to NWS.
- NWS has seemed very pedantic in specifying instruments for measurements.
- Operating personnel offshore do not see much encouraging evidence that data submitted are used by NWS or improve local forecasts in any perceptible way.

The working group concludes that the NWS and the offshore industry should make a concerted effort to reach accommodations and workable arrangements for submittal of critical weather data from offshore and then to enhance NWS utilization of such data.

There are many reasons to expect that near-term forecasting of fronts, squalls, or other storms that threaten offshore operating safety in the Gulf could be significantly improved with data from enhanced radar. The working group recommends that NWS and the offshore industry carefully discuss the deployment and most effective utilization of NEXRAD equipment at coastal locations adjacent to offshore operations. It may even be appropriate to investigate the merits of locating a NEXRAD unit on an offshore platform.

Aside from continuing production operations in the Gulf of Mexico and off southern California, there is little present activity in offshore areas adjacent to the United States. Seasonal exploratory drilling does continue on a very limited basis in northern waters off Alaska. The industry has recently taken new, unexplored lease positions in the Chukchi Sea, north of the Bering Strait. The pace of exploratory drilling and future production in these areas is uncertain. It is not realistic to make projections of forecasting needs for such operations at this time. Should the need arise for higher quality forecasts in these offshore areas, scarcity of weather data may be a real limitation. Development of all-weather satellite systems for weather data may be particularly critical for remote, northern data-scarce locations.

DREDGING

The dredging industry is subject to several classes of marine weather hazards. During mobilization to a worksite, a dredge may undertake a long distance ocean tow. Dredges are not particularly seaworthy and are very vulnerable to storm losses. In most respects, the towing of dredges resembles the towing of offshore jackup drilling barges. For both kinds of vessels, tows may take them through remote regions having very meager climatology for planning or forecasting for towing operations. Overall reduction of hazards is likely to come largely from upgrading worldwide forecasting.

During operations, dredges occupy a very specific site. The operating efficiency, safety, and damage risk for an operation depend very critically on reliability of forecasts. Dredges can be very vulnerable. Some units may be at risk when sea states exceed 3 to 4 feet significant height. The implementation of safety measures for most dredging operations is very much governed by local weather forecasts. To this extent dredges are like the inshore fishing fleet in need of meaningful improvement in the resolution of local forecasts.

For severe storms, safety and protection of the equipment calls for the dredge to be towed from the worksite to a sheltered anchorage. This takes a long time, 12 to 24 hours might be representative tow times. For some of the more exposed operating locations the tow time may be even longer and the vulnerability especially high. Longer range forecasts may be the only improvements in forecasting that would truly protect such operations. With that exception, it appears that dredging would benefit from most of the improvements for other coastal users.

Appendix G
Working Group 3:
Currents, Ocean Processes, and Ice

ALLAN R. ROBINSON, Harvard University, Leader
RICHARD B. ALLEN, Atlantic Offshore Fishermen's Association
PAUL H. GLAIBER, Great Lakes Dredge and Dock Company
WARREN W. HADER, Montauk Fishermen's Association
WALTER E. HANSON, International Ice Patrol, U.S. Coast Guard
SAUNDERS A. JONES, Puerto Rico Marine Management, Inc.
R. MICHAEL LAURS, National Marine Fisheries Service, National Oceanic and Atmospheric Administration
JAMES S. LYNCH, National Ocean Service, National Oceanic and Atmospheric Administration
CHRISTOPHER N. K. MOOERS, Institute of Naval Oceanography, U.S. Navy
DAVID F. PASKAUSKY, Research and Development Center, U.S. Coast Guard
WILLIAM C. PATZERT, Jet Propulsion Laboratory, National Aeronautics and Space Administration
DAVID J. H. PETERS, Conoco, Inc.
ALLEN M. REECE, Shell Development Company
GEORGE P. SPARACINO, Sun Transport Company
ANDREW M. SULLIVAN, Weather Network, Inc.
JOHN A. VERMERSCH, JR., Exxon Production Research Company

Working Group 3 was chartered to examine the requirements to improve observations and forecasts of oceanic currents and thermal structure, sea ice, and related fields (ocean structures, fronts, salinities, nutrients, pollutants, sound speed and propagation, and so on). An excellent cross-section of users (recreational, commercial, safety and enforcement, and military) and providers (academia, private sector, and government) participated in frank and open discussions.

Five specific topics were discussed at length.

1. What is the present status and perceived needs for nowcasts and forecasts of ocean currents, thermal structure, and related fields—globally and within the coastal ocean?

2. What is the prospectus and potential impacts of enhanced observations and forecasting capabilities?

3. What is the status of relevant ocean and ice models, observational networks, and forecast schemes.

4. What are the short- and long-term requirements for nowcasts and forecasts—accuracies, duration, frequency, and geographic coverage?

5. What are the present capabilities and future prospects for observing and forecasting sea ice, and what specific ice-related parameters are required?

Processes, such as waves, spray, and ice accretion, were the purview of other working groups. Furthermore, a number of issues were not examined (due primarily to the limited time available during the workshop and the limitations of the experience of the working group). These omissions included offshore mining applications, large-scale air-sea interaction and global change processes, planetary environmental management and pollution, and living marine resource issues (i.e., marine ecosystem modeling). Finally, sea ice was discussed and found to be an important issue, but the group's breadth of expertise was limited here.

CAPABILITIES

Ocean Observing Network

The inherent problem in monitoring and predicting the oceans has been the limitations of the observing network. Ocean observations have always been sparse, especially in areas seldom traveled by ships, and without a definable network except along coastlines. In addition, conventional ocean observation systems are very expensive to design, build, deploy, and maintain because of the harsh environment in which they dwell and the requirement of ship support time. Recent technological advances have allowed for a merger of in situ observations with remotely sensed data (land, ocean, and space based) and also for space-based communications systems.

TABLE G-1 Status of Operational and Planned Satellites

Satellite	Sensor	Velocity	Ice	Availability[c]
DMSP	SSM/I[a]		+	ongoing
NOAA	AVHRR	***[b]	+	ongoing
GEOSAT	Altimeter	+	+	1985-1990
ERS-1	Scatterometer	+[b]	+[d]	1990-1994
JERS-1	Synthetic aperture radar	***[b]	+[d]	1992-1995
TOPEX/POSEIDON		+	+	1992-1996

[a] SSM/I is probably underutilized
[b] Indicates velocity derived from flow visualization.
[c] Issues concerning availability
  1. Data acquisition will not be real-time without significant upgrades to processing and ground station capabilities.
  2. Uncertainties of data distribution.
  3. Lacking U.S. commitment to maintain a space-based ocean observing (altimeter/scatterometer) capability beyond GEOSAT.
[d] Ice from synthetic aperture radar, 1990-1995

Conventional platforms (such as coastal tide stations, moored buoys, drifting buoys, and ships) are more frequently incorporating sophisticated onboard electronic processing capabilities that permit the platform to provide some internal quality assurance and data storage capabilities, as well as to communicate via satellite to shore facilities and platform operators. Many of these systems are still expensive to deploy and maintain, but modern manufacturing techniques and reduced production costs due to mass production are becoming offsetting considerations.

Land- and ocean-based remote sensing capabilities are becoming important tools in monitoring physical properties of the ocean, especially in the coastal zone. Systems, such as the Coastal Ocean Dynamics Application Radar (CODAR) and acoustic Doppler current profilers, are emerging as cost-effective methods to determine ocean currents from shore, from oil and gas production rigs, from ships underway, and from sensors mounted on the seafloor. Microcomputers required to operationally process this information are becoming inexpensive, and can be packaged to be housed within the observing system.

Space-based remote sensing systems are becoming an essential component to the global ocean observing network. Two basic types of satellite sensors have been designed: passive systems, which receive but do not emit radiated energies; and active systems, which emit and receive radiated energies. Present status of operational and planned satellites is depicted in Table G-1.

Passive satellite systems include visible and infrared radiometers, imaging spectrometers, and passive microwave sensors. Ocean color instruments and imaging spectrometers can be used to measure the color of the ocean surface, and from this information they can imply the level of chlorophyll and primary productivity of the upper ocean, as well as the type of suspended sediments and sea surface temperature (SST) pattern. Passive microwave systems, such as the Special Sensor for Microwave/Imaging (SSM/I), can be used to delineate SST patterns, ocean heat fluxes, surface geostrophic currents, Gulf Stream shear waves, sea-ice boundary, concentration, type, and surface melt. Visible and infrared radiometers are capable of discriminating SST, ocean frontal boundaries, and sea ice boundaries.

Active satellite systems include scatterometers, altimeters, and synthetic aperture radars (SARs). Scatterometers and altimeters can be used to define wave heights, water levels and topography, ocean currents, surface wind stress, and sea ice location. SARs and other active microwave sensors can be used to characterize the structure of the ocean surface, including current and frontal boundaries, eddy fields and cold water regions, internal waves, bathymetric signatures, ocean wave spectra, and sea-ice extent, type, and motion.

Prediction Capabilities

From a predictive standpoint, considerable similarity exists between atmosphere and ocean, with mainly time and space scale differences due to density. The oceanic "internal weather systems" are smaller but slower than comparable atmospheric weather systems. Dynamicists see the ocean as a thin film, with a structure often reflected throughout the entire water column except for some submesoscale features and surface boundary layer effects.

From a dynamical view, all of the physical fields are coupled (interrelated). Even if we want to know a particular field, we may actually measure something else. Moreover, much of the forecast problem is similar to that of meteorology, complicated by turbulence and nonlinear processes, and characterized by a "limit to predictability." The kinds of ocean forecast problems include the following:

- Evolution via mesoscale internal dynamical processes, particularly for the energetic oceanic mesoscale that is analogous to the atmospheric synoptic scale. The oceanic mesoscale is generally remotely energized with mismatched coupling of the air and sea, resulting in long-term energy accumulation in the ocean.
- Evolution via local atmospheric forcing, fluxes of momentum (via wind stress), and fluxes of heat. These more rapid, forced, oceanic transients primarily occur in the upper ocean with atmospheric time scales.

Thus the oceanic forecast will then be ultimately limited by the ability to predict the atmosphere.

The requirements for observing the ocean for making forecasts are enormous, because of the small size of mesoscale "internal weather" features. Data requirements are very difficult to meet. Fortunately, most oceanic processes are somewhat slower than the atmosphere such that data can be assimilated over a longer period.

As oceanographers build their forecast schemes now, a modern approach to nowcasting and forecasting requires melding all available information into a total data set (satellite and in situ). Relationships between fields allow integration of temperature and velocity fields into a meaningful overall data set.

The SYSTEMATIC APPROACH (four-dimensional data assimilation) molds data sets and observations with dynamical model output. Dynamical interpolation allows for a three-dimensional analysis capability from a satellite data set. In situ observations can also be used for the analysis and nowcast. Such melded nowcasts are the most powerful. Continuous data assimilation permits the forecast to be updated and extends the validity of the forecast. The hope for useful nowcasts and forecasts lies with the SYSTEMATIC APPROACH, which optimizes information—thus providing the required resource for feasible and practical prediction.

The ocean prediction problem is often broken down into four scales: local scale (i.e., Chesapeake Bay), regional scale (i.e., the Gulf Stream Meander Region), basin-wide models (i.e., the North Pacific basin), and global models. Currently, the greatest emphasis within the Navy is on regional and basin-wide models, whereas efforts of NOAA have focused on local and global (climate) scales.

Global and basin-wide (dynamical) models will require considerable satellite data, and require state-of-the-art and advanced generations of supercomputers. The need definitely exists to maintain at least an altimeter in space until operational satellite systems are launched in the twenty-first century.

USER COMMUNITY REQUIREMENTS

A representative cross-section of the user community participated in identifying user requirements. Working group members represented the recreational and commercial fishing community, oil and gas exploration and production industries, marine transportation industries, the U.S. Coast Guard, and the U.S. Navy. A summary of user requirements is depicted in Tables G-2, G-3, and G-4.

TABLE G-2 Summary of User Requirements by Ocean Parameter (Excludes Sea Ice)

User	Ocean Temperature	Velocity	Location of Ocean Fronts	Forecast Frequency	Forecast Duration
Shipping industry	±1.0°C	±5 cm/sec	±6 nmi	1 per day	24-144 hr[a]
Fisheries industry	±0.5°C	±25 cm/sec	±6 nmi		0-48 hr[a], [b]
Oil industry		±25 cm/sec			24-72 hr
U.S. Coast Guard					
coastal	±0.5°C	±10 cm/sec	±1 nmi	4 per day	0-24 hr
offshore	±1.0°C	±20 cm/sec	±5 nmi	2 per day	0-48 hr
U.S. Navy	±0.25°C	±10 cm/sec	±5 nmi	4 per day	12-48 hr[c]

[a] Plus seasonal (3- to 6-month) outlooks.
[b] Plus 1-month planning outlooks.
[c] Plus 1-week forecast.

TABLE G-3 Summary of User Requirements[a] for Sea-Ice Information

Area	Primary Users	Resolution	Frequency	Duration
CONUS	Oil industry and aids to navigation	1 nmi	24 hr	0-6 hr
Great Lakes	Aids to navigation and flood management	1 nmi	24 hr	48 hr
Polar areas	Ship routing and pollution	10 nmi	12 hr	48 hr

Note: CONUS--Contiguous United States.

[a] Consensus of parameter requirements
  ice edge (location and movement, including leads and polynas)
  concentration of ice
  ice thickness
  ridging and related pressure fields
  development stages
  underice roughness

Marine Transportation Industry

The marine transportation industries that were represented included commercial shipping, barge owners, and coastal dredging. The user community uniformly requires information on surface velocity, and some cargo and tanker operations require nowcasts and forecasts of surface temperature.

Optimum ship routing requires accurate information of surface currents. Ship captains over the centuries have come to rely on sea surface temperature measurements and patterns as the only reliable tool to identify the location of the axis of the currents, but continue to rely on climatologies to estimate the speed. U.S. vessels operate throughout the world, and therefore require accurate current globally.

Cargo ships transporting perishable commodities (especially certain food items) and modern tankers carrying "exotic" fuels are beginning to require nowcasts and forecasts of ocean surface temperatures. Many of the commodities carried by these vessels are susceptible to warm or cold temperatures; in order to maintain thermal stability within the cargo and tanker holds, ship captains and navigators require accurate information.

Fisheries Industry

The fishing industry (both recreational and commercial) require a considerable variety of oceanic data. Fisheries management needs both physical and biochemical data and forecasts that affect eggs, larvae, and year class strength. Fisheries operations need accurate information to

TABLE G-4 User Requirements for Geographic Coverage

Shipping Industry
   Global requirements (predominantly along shipping lanes)
   Need highest resolution at major fronts and currents
      South Africa (AGULHAS)
      Southern Japan (KUROSHIO)
      Kuril Islands (OYASHIO)
      U.S. East Coast (Gulf Stream)
      Equatorial counter currents
      Selected straits and passages

Fisheries Industry
   Major, large, marine ecosystems
      Coastal: full water column over continental shelf
      Pelagic: full water column to base of thermocline

Oil Industry
   U.S. Exclusive Economic Zone
      Gulf of Mexico (as deep as 3,000 m)
      Continental shelf of the U.S. East Coast
   Foreign and international waters
      Off Japan
      West of the Shetlands
      Near Somalia
      Near the Amazon Delta
      Off northwestern Australia

U.S. Coast Guard
   Coastal and offshore waters of the contiguous United States
   Great Lakes
   Polar regions (Arctic and Antarctic)
   Other areas of the U.S. Exclusive Economic Zone

U.S. Navy
   Global requirements
   Regional (300 nmi x 300 nmi) support for moving battle groups

identify stock distribution, catch rates and quality, and other behavioral responses (feeding, depth, and geographic location).

The information is needed for a variety of "large marine ecosystems" that the U.S. industry manages or harvests. Many of these geographic areas are within the U.S. Exclusive Economic Zone (EEZ). Some, like the tropical Pacific tuna region, must be considered in a global context.

The fisheries community is rapidly becoming sophisticated in the type of information that it uses; such as temperature; salinity; dissolved oxygen; turbidity and light; chlorophyll, plankton, and carbon content; nutrients (nitrogen and phosphorus); pollutants; pH; transport and velocity; and color.

Oil and Gas Industry

The oil and gas industry, especially the "exploration" sector, requires accurate velocity nowcasts and forecasts in support of floating drilling operations to provide station keeping and for riser management. The industry representatives noted that accurately forecasting the size and track of loops, eddies, and meanders was extremely important for minimizing lost production time and platform damage. Accurate climatological data is needed to identify the maximum possible currents within their operating areas as use in risk management for permanent facilities (20- to 30-year life).

The U.S. industry operates in many parts of the globe. Within the U.S. EEZ, the industry operates in the Gulf of Mexico (as deep as 3,000 m) and along the continental shelf of the U.S. East Coast—representatives noted that the Alaskan North Slope will likely be inactive for an extended period. In addition, the industry operates in foreign and international waters off Japan, west of the Shetlands, near Somalia, near the Amazon Delta, and off northwestern Australia.

U.S. Coast Guard

The Coast Guard, as a user group, has five missions to which it needs accurate oceanic analyses, nowcasts, and forecasts. These activities include search and rescue, international ice patrol, marine environmental response (pollution), transit operations, and maritime defense. The majority of Coast Guard operations are within the U.S. EEZ, but the agency has a number of international responsibilities requiring global coverage.

Observations, nowcasts, and forecasts of surface currents for coastal and offshore areas are required to support search and rescue, marine environmental response, international ice patrol, and transit operations. Surface temperature information is required to support search and rescue, marine environmental response, and international ice patrol operations. Accurate nowcasts and forecasts of temperature, salinity, and turbidity for the entire water column are required to support the USCG's maritime defense mission (i.e., mine countermeasures).

U.S. Navy

Navy representatives concurred with the requirements outlined by the other users and included the need for accurate nowcasts of three-dimensional sound velocity fields. The oceanic information (0- to 48- hour forecasts) requires a global capability, but in limited areas (operational regions) at various times to support tactical oceanography operations and battlegroups in transit. Specific regional areas, based on maritime strategy and tactical requirements, include the North Pacific Ocean, North Atlantic Ocean, North Indian Ocean, Mediterranean Sea, and Norway and Greenland seas.

OTHER ISSUES

Processes

The working group came to a consensus that the understanding, observation, and prediction of a number of oceanic processes were fundamental to the safe and effective use of the ocean, such as

- internal ocean weather and related boundary processess, including upwelling, advection, frontal processes, mixed-layer processes, eddy-shelf interaction, and eddy shedding and meandering; and
- sea-ice processes, including ice edge (location and movement, leads, and polynas); sea-ice concentration and thickness; ice ridging, pressure fields, and development stages; and under-ice roughness.

COST-BENEFIT ANALYSIS

Each user group could identify meaningful and sustainable benefits from improved observations and forecasts of oceanic phenomena. Table G-5 highlights the basic categories of benefits for the user communities.

SUMMARY

FINDING: There exists a common national interest in, and need for, nowcasting and forecasting oceanic velocity, temperature, sea ice, and related fields. Significant and sustainable benefits to a variety of commercial, military, and recreational oceanic activities are identifiable based on existing national capabilities.

Recent progress in understanding oceanic processes, as well as new and innovative observing methods (conventional in situ and remote sensing systems) and forecasting techniques, indicate the nation's technological and scientific readiness to provide operational oceanic nowcasts and forecasts. Short- and long-term requirements for oceanic observations and forecasts have been identified by user groups, and include specific parameters, geographic location, duration and timeliness, and accuracies.

RECOMMENDATION: An increased national effort is needed to improve or establish operational capabilities that should include

TABLE G-5 Cost-Benefit Relationships, by User Community, of Improved Ocean Observations and Forecasts

Cost-Benefit Category	Marine Transportation	Fisheries	Oil and Gas	Coast Guard	Navy
Minimize ship and aircraft time	+				
Increased fuel savings	++	+		+	++
Increased shipkeeping efficiency	+		++	++	
Reduced cargo damage		+++			
Increased harvest		++			
Minimum lost production time			++		
Reduced equipment damage	+			+++	+++
Increased potential lives saved				++	
Increased property recovery			+		
Improved environmental response					+
Drastic savings on use of SONABUOYS					

- expanded and accelerated efforts to (1) produce fisheries forecasts and (2) develop an observing and forecasting capability to support commercial, military, recreational, and research activities within the U.S. Exclusive Economic Zone (EEZ);
- coordination and cooperation of ongoing and component efforts of federal and state agencies, academia, and the private sector; and
- a national policy on the importance of the oceans, especially the EEZ, to the welfare and economic prosperity of the nation.

Appendix H
Working Group 4: Nearshore Forecasting

WILLIAM G. GORDON, New Jersey Marine Sciences Consortium, Leader
PAUL H. GLAIBER, Great Lakes Dredge and Dock Company
GARY L. GRIDLEY, Conoco, Inc.
WARREN W. HADER, Montauk Fishermen's Association
WALTER KRISTIANSEN, Amoco Transportation Company
KATHLEEN A. MILLER, National Center for Atmospheric Research
STEPHEN K. RINARD, National Weather Service Southern Region, National Oceanic and Atmospheric Administration
CHARLES L. VINCENT, Coastal Engineering Research Center

Today more than ever, U.S. citizens depend on the coastal area to support a variety of economic activities. In the near future 75 percent of the U.S. population will live within 50 miles of a coastline. Increasing users of the area expect and will need timely observations and forecasts of weather, oceanic, Great Lakes, and river phenomena, such as weather, wind, waves, ice and icing conditions, temperature, current, and nearshore processes. Benefits will accrue in safety, economic efficiencies, and rational development. This report of Working Group 4 defines the range of user groups, establishes their needs for information, reviews present capabilities, and suggests areas for improvement of near-term and future operation and research to satisfy each need. For the purpose of this workshop, nearshore includes tidal to 200 nautical miles (nmi) offshore, the Great Lakes, and other inland waterways.

Coastal waters, with their complex water, land, and air interactions, provide most of our fishery resources. The relatively shallow areas of the marine complex contain energy resources important for the U.S. economy. Increasingly the same areas are used for direct disposal of a variety of unwanted materials. A wide variety of large oceangoing craft, (military and civilian), commercial and recreational fishing vessels, and recreational craft numbering into the millions transit or depend exclusively on these areas. The national economic use of this coastal area requires short- and long-term forecasts of weather and oceanic conditions to assist in protecting such areas from unwanted abuse and destruction, for safe conduct of activities, and for rational development.

PRESENT MARINE WEATHER SERVICE

The majority of nearshore mariners receive weather information from various sources. Most recreational boaters for example obtain weather information from commercial radio and television stations or from the Weather Radio of the National Oceanic and Atmospheric Administration (NOAA). These offer good service for those operating at short distances offshore. A common complaint about NOAA Weather Radio is that of limited range (line of sight for very high frequency [VHF]). Commercial radio and television offers land coverage, but frequently such coverage is inappropriate over water because it does not take into consideraiton the substantial variation that can occur over water.

Weatherfax machines on board vessels allow for the receipt of many different types of weather and oceanographic maps, but few recreational craft and only the larger commercial fishing vessels are equipped with them. Radio facsimile broadcasts are directed to the high seas and are often poorly received nearshore due to skip of the signal. Satellite images of oceanographic features (e.g., warm-core rings, eddies, temperature, and current boundaries) are used to evaluate prime fishing areas. As an example, satellite pictures of the Gulf Stream, depicting temperature boundaries and eddies, are routinely used by fishermen in locating swordfish and other large pelagic species.

Coast Guard radio also services such craft with notice to mariners and relays National Weather Service (NWS) advice on weather and wave conditions. Larger oceangoing craft depend to a greater extent on commercial services and ocean conditions.

CURRENT MARINE OBSERVATIONS

At the national level, observations on marine weather and oceanic conditions are collected by the NWS of NOAA. These include buoys, C-Man, Volunteer Ship Observations (VOS) program and the Marine Reports Network (MAREPS).

The buoy program is managed by the National Data Buoy Office (NDBO). The buoys measure basic meteorological factors such as wind, temperature (air and sea), pressure, and waves. These buoys report once an hour into NWS communications via a GOES satellite. They are located at 54 coastal and high-seas sites. The main advantage is that they are obtaining over-the-water accurate and reliable data and provide good time series and climate data. Some buoys report spectral information, and a few of those also report wave direction.

C-Man, or the Coastal observation network, is operated by the NDBO in a similar fashion as the buoys, except this system does not normally observe waves. Most sites have very good marine exposure but are representative of a coastal location and are not over the water such as the buoy system.

Through VOS, offshore ship operators reports are sent via CW to land-based radio stations (Coast Guard and commercial) and then into NWS. More modern methods also employ satellite communications. Most observations are subjective (except pressure). This forms the basis of all high-seas observations.

The marine environment is a data environment—very few observations are available over broad ocean areas. The MAREP network uses marinas, Sea Grant, and ham and contract radio operators to collect uncoded marine observations from small boats on position, wind, seas, visibility, and weather. These observations are in turn relayed to NWS marine forecasts offices. Such observations assist the marine forecasting and verification programs with the result being more accurate forecasts and warnings and a safer marine environment. A more immediate user payback is the anticipation of direct user input into the forecasting program and the ability to obtain the latest marine forecast and warning through the MAREP operator.

Observations from the above as well as from satellite and land-based sources are utilized to develop the present array of marine weather and ocean condition forecasts.

MARINE NEARSHORE USERS

A variety of users need and use nearshore ocean and weather observations and forecasts. All users make use of marine and general forecasts in their day-to-day activities; during critical weather periods this use increases. Such users are identified in Table H-1, along with needs, frequency of need, and potential areas for improvement.

TABLE H-1 Nearshore Forecasting Users, Needs, and Improvements

User	Needs	Frequency of Needs	Needs Improvement
Commercial fishermen	Wind, wave temp., tides	6 hours	More site specific
Recreational fishermen	Wind, wave, weather	6 hours	More site specific
Recreational boaters	Wind, wave, weather	6 hours	More site specific
Beach goers	Wind, wave, weather, Tide	12 hours	More site specific
Coastal construction	Wind, wave, weather tide	12 hours	More site specific
Tug and barge	Wind, weather, tide	6 hours	Increased reliability
Marine enforcement	Wind, wave, weather	6 hours	Increased reliability
Oil & gas development	Wind, weather	12-24 hours	Increased reliability
Disaster preparedness	Wind, weather	12-24 hours	Increased reliability & precision
Distant water merchant	Wind, wave, tide	6 hours	Very site specific

USER NEEDS

Although there is a significant level of satisfaction regarding current services, various user groups identified areas where improvements in services would enhance their activities.

In general the greatest need to National Weather Service operations is to achieve improvements in observation capability. Better, more comprehensive observations are the prerequisite for improvements in short- and long-term forecasts. Of nearly equal importance is the need to improve the quality of dissemination. A large number of users depend on the present system of voice transmission, and the present NOAA capability is limited by range was well as geographic coverage. Improvements in this aspect alone would substantially improve service to many users of the nation's coastal

TABLE H-2 Inadequacies in Marine Observation Capabilities

Technique	Number Displayed	Level of Adequacy	Comments
Radar		Absolute-NEXRAD deployment under way	By 1995, system fully operational
Ocean buoys	54	Highly accurate, and reliable; present coverage inadequate for observation of marine weather and oceanic data	Costly to install and maintain; good source of data
C-Man	41	Coverage incomplete; dependable, reliable hourly observation	Dependent on coverage; results appear cost effective
Volunteer ship observers		Data highly subjective and confined to ship routes	Forms basis for all high-seas observations
Marine reports	20 stations	Incomplete coverage, very limited to radio coverage at station location	Provides for high level of cooperative efforts

waters. Geographic coverage was also a matter of considerable concern. Presently, weather forecasts are issued for a fairly broad area along the coast or within an estuary, such as the Chesapeake Bay. Users of these areas, particularly the recreational fishermen and boater and the small-scale commercial fishermen, find that such forecasts do not offer small enough resolution. Improvements to a large measure depend on increases in the number of observations.

ADEQUACY OF PRESENT CAPABILITIES

As noted above, current capabilities of marine observations are inadequate to provide for significant improvement in the current level of forecasts. These inadequacies are summarized in Table H-2.

Inadequate support results in significant losses to users. Losses occur in a variety of ways and are not fully quantifiable at this time. Improvements in coverage, level of accuracy, reliability, and so on will to a large degree offset those losses and include multiple groups.

The benefits will include cost savings to commercial operations that can modify their operations in response to the better forecasts. In many cases there will be multiple potential beneficiaries. The expected cost savings per period for the entire set of these commercial operators would be relevant. Therefore the benefits are likely to be highest in heavily used shipping lanes, congested terminal areas, areas of high recreational use (boaters, beach goers), and commercial fisheries. Other commercial benefits include improved safety, potentially measurable as reduced loss of life and expected dollar value of reduced equipment damage.

One important and often neglected point is that not all of the benefits of improved forecasts will accrue directly to the commercial operators making use of the forecasts. In highly competitive industries the benefits may also accrue to the consumers. There may also be benefits in the form of reduced risks of environmental damage from oil spills and so on.

To the extent that the expected benefits are broadly diffused across a wide spectrum of individuals, private forecasting services might be unable to provide the improved service even though the potential new social benefits might be large. In that instance, public provision would be more efficient. Public provision would also tend to be more efficient in cases where it is difficult to exclude nonpayers from using the service (e.g., information transmitted via radio). Benefits will accrue to recreational boaters and commercial operators who support sports fishermen. One clear immediate benefit will be improved safety. In addition to the direct benefit of reduced loss of life and property, there is the indirect benefit of reduced public expenditures on search and rescue. In addition to improved safety, improved quality of recreational experience should be identified. If better forecasts allow recreationists to make better decisions about activities on any given day, they will enjoy a higher average quality of recreation. Such benefits would be difficult but not impossible to quantify. Techniques are now available to estimate the value of quality attributes of recreational sites. Similar techniques could be applied to this problem.

IMPROVING THE PRODUCT AND SERVICES

The user community needs to be better informed about the governmental weather community's capabilities. Many users are not fully aware of what the government agencies can provide. Thus, a program could be developed to educate the user community about available government data, programs, and other services. The community could then access existing information and take better advantage of it. Also, as new or improved programs are developed by the government, the availability and capabilities of these programs should be made readily available for public knowledge.

The general concept of improving technical forecasting has considerable merit. However, for it to be useful to the public and for it to remain within the realm of NWS responsibility, it is apparent that serious study will be required. Care must be exercised so that one particular user group does not appear to be gaining the majority of the derived benefits. The NWS depends on observations of weather and ocean conditions. Many private industries can accurately make such observations. NWS and these groups should try to develop mechanisms to update the absorption of this information into the system. In most cases, private industry would respond to this concern because better forecasts would be to their benefit. In other words, the user community would be more involved by gathering information for the forecasting program. The foremost unmet need is enhancement of aquatic environment observations. In general, observations over inland areas (i.e., Great Lakes and inland waterways) are considered adequate for production of appropriate forecasts. Over the coastal and ocean areas, significant improvements are required to forecast weather oceanic phenomena. Installation of equipment such as radar (NEXRAD) and improved computer capacity should result in very significant improvements in marine weather forecasting capability. Other cooperative efforts such as C-Man and MAREP can result in significant improvements in coastal area weather observations. Additional steps should be explored to involve industry in data collection.

Obtaining additional oceanic observations will require significant expanded efforts and coordination. Cooperative efforts such as those mentioned above can readily be expanded. Some would require minimum costs, while others could be extremely expensive.

RECOMMENDATIONS

Within present resource conditions, the working group recommends

- improved observation through greater cooperative arrangements, such as C-Man, MAREP, and others;
- improved dissemination through changes in format, area of coverage, and adjustments to radio transmission;
- improved public understanding of weather service capabilities, increased public awareness of marine users, and training of weather observers; and
- refocusing on user needs through discussions with various user groups.

With limited additional resources, the working group recommends

- improved observations as noted above and adding ocean buoys that are properly placed; and

- terminal forecasts be provided for entrances to major harbors and rivers.

With significant additional resources, the working group recommends

- additional ocean buoys;
- satellites;
- expanded cooperative arrangements; and
- at-sea ship radios and training of ship personnel.

In the area of surprise storms, the working group recommends

- continued research on physics,
- continued development of empirical techniques,
- research that includes observations involving satellite and forecasting techniques,
- issuance of appropriate warnings,
- user education,
- institution of three-hour reporting.

Recommendations Concerning Overall Capabilities

Increased effort is needed to improve or establish operational capabilities for nowcasts and forecasts of velocity, temperature, and related fields within the coastal and deep offshore ocean, including fisheries forecasts and the development of an EEZ capability. Coordination and cooperation of ongoing and component efforts is necessary, as is the development of a national policy for internal ocean weather nowcasting and forecasting.

There is an urgent need to establish a national ocean monitoring satellite program for the routine operational observation of marine parameters (wind, waves, sea-surface topography, sea color, temperature).

Recommendations for Dissemination Systems

The existing services of high frequency facsimile and radioteletype should be maintained. Satellite transmissions should be improved and broadened.

NOAA Weather Radio should be upgraded to provide scheduled broadcasts (i.e., after-the-hour marine forecasts at 27 nmi). Its range of broadcast and resolution of marine forecast should be improved.

NAVTEX broadcasts should be more frequent than once every 6 hours and warnings should be disseminated by all media as they are issued. Adequate time allocations for broadcasts should be ensured.

HIGHLIGHTS

The best service the NWS could provide the mariner is the most detailed, most accurate forecast possible, out as far in time as possible. Accurate detailed forecasts out 5 days would be extensively used.

There is a need for a "marine terminal forecast" outside the entrance to major harbors and rivers. Such a forecast would be widely used for planning purposes by mariners entering and exiting port.

CW and radio teletype communications are being phased out in favor of satellite communications. NWS should emphasize distribution of weather information and observation collection using satellites.

Radio facsimile weather broadcasts are often not well received near-shore. Since many users operate near the coast and are increasingly equipped with facsimile receivers, a program needs to be developed that would ensure that this high density of users is not omitted from radio facsimile reception.

To recognize ship participation in the cooperative ship observation program, NWS should provide a pennant or some type of recognition that could be displayed onboard ship.

Most shipping companies would welcome NWS forecasters to sail on their ships to familiarize the forecaster with shipboard operations and the crew with NWS procedures.

Appendix I
Working Group 5:
Collection, Reporting, Dissemination, and Display

RICHARD WAGONER, National Weather Service, National Oceanic and Atmospheric Administration, Leader
RICHARD B. ALLEN, Atlantic Offshore Fishermen's Association
STEVE COOK, National Ocean Service, National Ocean and Atmospheric Administration
HENRY CHEN, Ocean Systems, Inc.
GLENN D. HAMILTON, National Data Buoy Center, National Oceanic and Atmospheric Administration
ROBERT E. HARING, Exxon Production Research Company
PAUL B. MENTZ, Maritime Administration
ROBERT J. MURRAY, Matson Navigation Company
KENNETH W. RUGGLES, Systems West, Inc.
PAUL E. VERSOWSKY, Chevron U.S.A., Inc.
JAMES W. WINCHESTER, Association of Private Weather Related Companies
VINCENT ZEGOWITZ, National Weather Service, National Oceanic and Atmospheric Administration

The working group reviewed all aspects of data collection, reporting procedures, methods of dissemination, and product format and display. In terms of overall emphasis, the categories of data collection (especially from ships at sea) and dissemination were discussed more than the other two categories. The general feeling was that reporting procedures were mostly adequate. Format and display as a category was considered important but not a large issue in that current and future technologies would naturally lead to more enhanced displays and that user groups had been successful in making their needs known to providers of weather information. Recommended actions are summarized on pages 124-125.

COLLECTION OF MARINE OBSERVATIONS

The working group identified the following deficiencies that contributed to a serious degradation of marine observations and weather data collection from the marine environment:

- significant loss of marine data within complex communications systems (within and external to the National Oceanic and Atmospheric Administration [NOAA]);
- significant delay of critical weather data delivery to forecasters;
- insufficient provision for communicating valuable, unused data sets to weather forecast providers;
- antiquated, slow communications systems in many cases; and
- insufficient use of reliable, quality-controlled satellite communications for data collection.

Delayed and Lost Data

Ship reports comprise continuing and vital input to the marine synoptic forecasting process. Since the total input to the marine forecasting system consists of buoy, ship, and satellite data, the importance placed on the ground truth data provided by ships cannot be overstated.

There are a great diversity of communications methods available for delivery of these marine reports that have developed in the last 50 or 60 years, such as CW, voice, radio-teletype, and satellite. In the process of passing through these systems, a significant amount of marine data has been determined to be lost. Causes for data loss are widespread and not attributable to a single source of error but include

- noncompliance with operational procedures,
- software deficiencies,
- outmoded communications procedures, and
- the diversity of delivery methods and formats

This amount of data loss is estimated at 25 to 30 percent, representing a loss of 50,000 to 60,000 marine meteorological messages per year. An in-depth examination of the marine data delivery system should be undertaken with the objective of minimizing the current and long ongoing data loss.

Feedback mechanisms should be instituted to provide the data gatherers and communications personnel with information on the effectiveness of their actions.

Concurrent with the data loss problem is the issue of nontimely marine data receipt. Since synoptic marine data is highly perishable, it is of great importance to deliver these data to the marine model operator model in time to be used in the first run issued for forecast guidance (1 hour). Approximately 45 percent of the data currently received arrives too late for inclusion in the first-cut synoptic forecasting model run and is merely archived. Streamlining communications procedures would not only mitigate the data loss problem, but would serve to increase the percentage of usable data available for improved synoptic forecasting.

User marine forecasting needs have indicated that data collection efforts, particularly quantity, will need to increase. Unfortunately, perishable weather data must compete often with a myriad of messages handled over government and private networks. To facilitate timely receipt of data, marine voice and radiotelegraphy channels should have identified periods at synoptic hours 00Z and 12Z, and if possible, 06Z and 18Z, that are reserved for weather reporting. This may require a period of up to 30 minutes wherein only weather reports are handled by marine radio stations. This policy serves not only to facilitate data transmittal, but to emphasize to the communicators the importance of forwarding data to the National Weather Service (NWS).

Unused Sources of Data

The working group expressed concern regarding many sources of valuable marine data that are underutilized. The reasons for not using these data sets included insufficient communication systems for relaying data to weather providers, proprietary issues, and in some cases government regulations regarding collection, protocols, and quality control.

Presently the U.S. Coast Guard is in receipt of many ship reports to the AMVER program that are treated as proprietary information and not released to any external users. This represents a significant resource of potential data for input to numerical analyses. The working group recommends that the AMVER reporting system be changed so that the reporting ships must also report data required in the standard WMD observation (if not already required) and transmission of these reports to the Numerical Meteorological Center (NMC) with the call signs changed to SHIP so as to maintain the proprietary concern of specific shipping firms.

There are other data sets available from private firms who are willing to contribute them to government providers if an economical and reliable method of communication is established. In some cases government providers like NOAA are unwilling to accept these data because of concern about quality control and data protocol. These are understandable issues, however, the working group feels that a method and system should be developed to acquire, validate, and assimilate near-real-time observations of sea surface, atmospheric, and subsurface parameters from volunteering users on an intermittent basis. There is a large, untapped source of valuable data that is not regularly reported (as in voluntary observing ships and SEAS) or transmitted from permanently occupied sites (as in NDBO buoys). These sources could be, for example,

- mobile drilling units (MODUs) in a remote area for a period of weeks to months at a specific site;
- transocean movement (tows) of MODUs, dredges, and barge-mounted fabricated structures;
- proprietary environmental monitoring and system response programs where basic atmospheric data can be "declassified"; and
- existing permanent oil and gas platforms located in data-sparse areas (widely practiced in the North Sea).

To accomplish these goals will require cooperation between NOAA's forecasters and the private sector, as well as preplanning and flexibility. It is mandatory that NOAA has a clearly identified procedure and responsible group to respond quickly to the availability of such data from these private, temporary sources. The procedure should be flexible in terms of format, qualification of data, and timing. The private sector should be able to make a phone call, say they would like to donate certain data, and be assured it will be received and appropriately utilized.

The working group felt that some components of the complex communication systems used for collection of marine data should be significantly upgraded or replaced. Special attention was drawn to satellite communications as a streamlined, timely, and highly reliable method of communicating weather data.

Satellite-based technology like the SEAS data collection system was viewed favorably by the working group. The members of the working group felt that the committee should strongly recommend a cooperative initiative between the government and private shipping firms to increase the deployment of SEAS units over the oceans and coastal water.

Expected Improvements and Benefits

Improvements in the data collection system will result in many improvements in the products and services provided by both private and public entities. The working group attempted to identify some of the significant areas of improvements and benefits.

Numerical analysis, the bread and butter guidance for all marine forecasting (both public and private), will almost certainly be improved as a result of recommended changes in the data collection system. More data at the two primary levels (surface and commercial jet level) will result in more accurate initialization of location and intensity of major low-pressure systems, jet stream location and intensity, and better definition of ocean conditions like sea surface temperature. This increased accuracy will translate into more accurate and reliable analyses and numerical prognoses.

This impacts directly on container vessel operations in the offshore service; the following benefits are realized. The key to a successful inter-model transportation system is in rapid but dependable schedule keeping. The marine section of the transportation system relies on the vessel's ability to maintain the required speed to perform the given schedule. The chosen route depends on the accuracy of the weather forecast, which will, in turn, produce the optimum route to follow. A principal cost item in ship operations is fuel consumption. Fuel usage is dramatically lessened by choice of the optimum route. In addition, hull and ship damage costs can be considerably reduced by accurate forecasting of wave heights and direction. The planning of deployments, including scheduling planning, depends on accurate history of weather to be experienced in the area to be traversed and how vessels will perform in that area. This depends on the quality of the historical data available.

Another important result of enhanced observational data (and their effective dissemination) will be increased confidence in area forecasts and resolution of local disparity between observation and forecast. By necessity the model results (forecast) will be averaged over a relatively large area of open ocean. (This may be mitigated nearshore when NEXRAD becomes operational.) Local users will often find marked differences in conditions within the forecast area. Direct access to nearby observational data will permit the user to validate his judgment and more properly interpret the forecast for his purposes. The dissemination of all regional observations will supplement the forecasted conditions.

There are many users who operate permanent facilities offshore. Severe weather events like hurricanes and intense low-pressure systems in higher latitudes may require offshore operation to curtail production or completely execute the facilities. Improvements in forecasts as a result of enhanced collection of data will allow operators to minimize losses and provide the maximum level of safety for personnel.

In general, offshore operators establish weather criteria for the curtailment of operations or evacuation of platforms. In both cases there is a cost involved, chiefly lost production (barrels of oil and gas) and lost productivity.

PROBLEM: Cry wolf—incorrect or inaccurate forecasts that shut down operations need to be eliminated.

- **PROBLEM:** Fear that weather will close in before evacuation can be achieved. (Lack of knowledge of weather problems of something other than tropical cyclone.)
- **PROBLEM:** Forecasts that do not agree with measured observations of the user. Winter storms, the passage of fronts, and post-hurricane return events pose particular problems in this regard.

The cost for 1 day of oil production loss for a company producing 100,000 barrels of oil per day at $16/barrel is $1.6 million. Add to this the fixed cost of operations. If improved data collection and dissemination is achieved, then we can get better model results and forecasts from the National Weather Service and better forecasts from private forecasters.

DISSEMINATION OF MARINE WEATHER PRODUCTS

The working group spent considerable time discussing issues related to the dissemination of weather products to the marine user. The following general points were made.

- A significant level of user education is required regarding the availability of existing services offered by both public and private providers.
- Users in the congested nearshore zone (coast to about 100 miles) require enhanced dissemination systems.
- The growing numbers of users in the 50- to 200-mile zone from shore require enhanced dissemination systems.
- All users require timely dissemination of products with a standard format where possible.
- There is a growing subset of users who require fast, error-checked, electronic dissemination and display of data, including gridpoint fields for local processing.
- The cost of dissemination (primarily INMARSAT) is a major concern for weather providers (public and private) and the user.

User Education and Enhanced Suite of Data Available to Users

Considerable data can be available to the user that can assist in the decision-making process. NOAA and NWS should attempt to define the user market and provide the various markets with a list of available data, along with the format and method of retrieving that data.

Once the markets are defined the agencies should advertise the available products and disseminate information to those markets through associations, educational institutions, power squadrons, and special meetings advertised and conducted by government agencies.

Shipping association members and government agencies should promote meteorological and oceanographic courses that include real-world product availability, weather and communication systems, waves and lull response, and techniques of estimating or measuring weather phenomena.

Nearshore User

From the shore out to about 100 miles, small boats and commercial operations require a more comprehensive weather data and warning service. In this area the NOAA Weather Radio's (NWR) primary means of communicating weather information and warnings is considered inadequate in its present form for nearshore support. Users require a printed medium for forecasts, data, and warnings because of busy and often noisy working environments. The new Coast Guard NAVTEX system may satisfy these needs. Alternatively, upgrading of the NWR may be necessary.

Offshore User

Beyond 50 nmi and within 200 nmi the private recreational boater and fisherman should be expected to invest in a NAVTEX receiver to ensure the safety of his or her vessel.

The International Maritime Organization will amend the Safety of Life at Sea (SOLAS) Convention in November 1988 to require NAVTEX for ships 300 tons or larger. The broadcasts for NAVTEX should be for the 50 to 200 nmi offshore weather. Very high frequency (VHF) voice transmission should provide forecasts for nearshore weather out to 50 nmi.

The exploration of the U.S. EEZ by the public is expected to increase as marine forecasts improve. As the public moves farther offshore, standards for safe boating should be expected to increase to reduce the potential for distress. Marine forecasts must be made available to this sector of the public. Dissemination will be most efficiently handled by NAVTEX. NAVTEX therefore should be used as a marine receiver required for safe operation seaward of 50 nmi.

Digital Data Transmission

There is a growing population of users who require fast, error-checked transmission of digital weather data for local processing. Most users want these data in grid-point format.

The most cost-effective way of transmitting digital weather information is via the satellite communications network (INMARSAT represented by COMSAT in the United States). The guided data can then be used to generate weather charts as well as create a data base for voyage planning, damage avoidance, and so on.

Due to the compressed nature of the data formats and the noise level of the SATCOM, the transmission must be error checked and verified. Commercially available software and hardware are available to make sure the transmission is error free.

SATCOM stations are more affordable ($30,000–$50,000), compact, and light. The rates have been reducing as more users get on-line. There are more than 6,000 ships worldwide that are equipped with SATCOM terminals. The transmission cost is less than $10/minute.

Cost of Dissemination Channels

Many users and providers represented in the working group indicated concern about the cost of both collection and dissemination of weather-related information via INMARSAT. Although the general feeling was that these costs would gradually decline, the fact that INMARSAT was the only viable nongovernment, satellite-based communications channel available causes a difficult situation for value-added companies and users.

Product and data dissemination to coastal and high-seas marine operators presents special communications issues that are not now effectively addressed by existing public and private services. On the high seas, traditional high-frequency, full-period radioteletype and facsimile broadcasts for the high seas are being discontinued or curtailed. These services are vital to the safety and economic efficiency of high-seas marine operators, but are not being replaced by equivalent or better services. The government should develop a plan in cooperation with the marine and value-added weather community to restore an effective level of service to the high-seas marine community.

RECOMMENDATIONS

1. All reporting platforms (fixed and floating) use automated satellite data collection and transmission systems whenever possible. Automated collection spans the range from present "SEAS" type to fully automated systems developed in partnership with private industry. A national policy advocating this technology should be advanced with the goal of outfitting all U.S. volunteer observing ship platforms.

2. Existing shipboard computer equipment and INMARSAT communications equipment should be utilized for preparing and transmitting routine voluntary weather observation reporting wherever and whenever possible to improve the accuracy and timeliness of such reporting from commercial maritime vessels.

3. The U.S. Department of Transportation/Maritime Administration and the U.S. Department of Commerce/National Oceanic and Atmospheric Administration should obtain appropriate authorizations of and appropriations for research and development and technology assessment over a 5-year period to further develop public and private sector initiatives to improve the collection, reporting, dissemination, and display of global weather and ocean dynamic forecasts for the maritime industries of the United States to enhance maritime safety, productivity, and competitiveness.

4. A comprehensive education and training program should be initiated by the U.S. maritime industries to ensure that navigation (deck) officers of commercial maritime vessels are fully aware of the availability and utilization of global weather and ocean dynamic forecasts provided by both public and private sector services, and to ensure that they are sensitive to the need for voluntary weather observation reporting to support the development of such forecasts.

5. The government should dedicate selected communication channels to high-priority data collection at synoptic hours 00Z and 12Z, so as to avoid delays from competing messages.

6. Beyond 50 nmi and within 200 nmi, the private recreational boater and fisherman should be directed to acquire a NAVTEX receiver to ensure the safety of his or her vessel, rather than rely solely on VHF transmissions.

7. The government should plan to develop in partnership with the marine and value-added weather providers, a system for the provision of currently required weather products and future data products that are anticipated.

8. NOAA should upgrade its NOAA Weather Radio dissemination system in areas where it presently operates.